Milk and Meat
From Grass

J.M. Wilkinson
B.Sc., Ph.D., M.I. Biol.

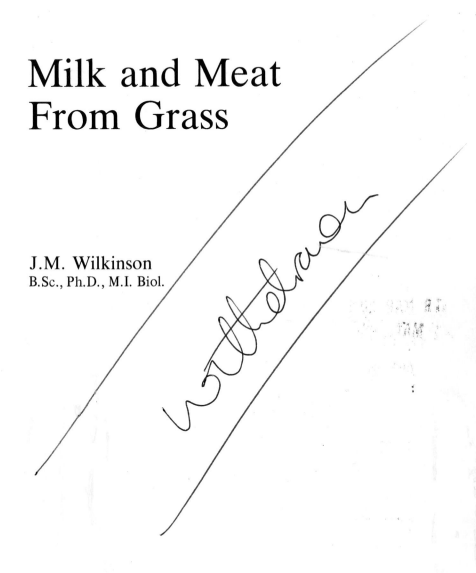

GRANADA
London Toronto Sydney New York

Granada Technical Books
Granada Publishing Ltd
8 Grafton Street, London W1X 3LA

First published in Great Britain by
Granada Publishing 1984

British Library Cataloguing in Publication Data
Wilkinson, J.M.
  Milk and meat from grass.
  1. Grasses   2. Feeds
  I. Title
  636.08'55   SF99.G7

ISBN 0–246–12290–0

Typeset by Cambrian Typesetters
Frimley, Surrey
Printed and bound in Great Britain by
Mackays of Chatham, Kent

Contents

Preface

This is a book about converting grass, the most important agricultural crop, into milk and meat for human consumption. It focusses on what can be achieved on commercial farms as well as on information from recent grassland research and development work.

Profit from livestock on grass means reducing wastage by matching the quality of grass and its availability to the needs of the animal at all times. It also means paying attention to reducing losses during conservation of grass crops as silage or as hay. Techniques for upgrading low quality crops during their conservation are also outlined.

The potential for making more efficient use of the valuable resource of grass in temperate climates is enormous. Even the enthusiastic grassland farmer will find room for improvement.

The economic climate for producing milk and meat from grass is likely to become relatively less attractive during the 1980s as controls are applied to curb over-production in the European Community. Thus the value of grass as a relatively low-cost feed for ruminant animals is likely to be more generally appreciated. In areas where grass grows well, this means an increased contribution by grass to the total feed requirements of the animal.

The main emphasis in this book is on the efficient management of the grass crop, particularly when it is grazed and conserved for the winter period. Attention is also paid to integrating legumes into grassland farming, especially in lamb production. The highest margins are achieved from high animal performance at relatively low input costs.

I am grateful to Mr Denis Chamberlain, Editor of *Farmers Weekly*, and to Mr Peter Jones, Associate Editor, for their assistance in the preparation of the book, which is based on a series of articles commissioned by *Farmers Weekly* as part of the 'Money from Grass '83' Campaign. The articles were published in *Farmers Weekly* between February and December 1983. Many of the drawings and the photographs are reproduced courtesy of *Farmers Weekly*.

My wife, Cherol, gave me both moral and physical support throughout the project. Without her help and encouragement the many deadlines would have been missed. She typed the manuscripts late into the night, without complaint and with very few errors. Those that appear here are my fault, not hers.

J.M. Wilkinson

Abbreviations

ADAS	Agricultural Development and Advisory Service
BGS	British Grassland Society
BRA	Beef Recording Association
DM	Dry-matter
EHF	Experimental Husbandry Farm
GJ	Gigajoule (a thousand megajoules)
GRI	Grassland Research Institute
ha	Hectare (2.471 acres)
HMSO	Her Majesty's Stationery Office
ICI	Imperial Chemical Industries plc
MAFF	Ministry of Agriculture Fisheries and Food
ME	Metabolisable Energy
MLC	Meat and Livestock Commission
MMB	Milk Marketing Board
MJ	Megajoule (a thousand joules)
N	Nitrogen
NH_3-N	Ammonia nitrogen, usually expressed as a percentage of total nitrogen
NIAE	National Institute for Agricultural Engineering
pH	A measure of acidity and alkalinity; 7 is neutral, above 7 is alkaline and below 7 is acid
UDP	Undegraded dietary protein
UKASTA	United Kingdom Agricultural Supply Trade Association
UME	Utilised Metabolisable Energy

Conversion

To convert kg/hectare into units/acre, multiply by 0.8.

1 Components of Success

Success and how to achieve it

Success, in this book, means making money from the efficient use of land, capital and labour to produce milk, beef or lamb. Some farmers are more successful than others. Why? Are they better traders, technically more proficient or just fortunate in having plenty of good land, capital and labour?

Studies in the UK of recorded milk, beef and lamb producers by the Milk Marketing Board (MMB) and the Meat and Livestock Commission (MLC) show clearly that financial success as measured by gross margins, is closely linked to technical performance. This indicates that in beef and lamb production, the way to higher gross margins is through improved physical performance as much as through being a good buyer and seller of stock.

In dairying, the MMB has taken the analysis right through to profit, management and investment income, and return on tenant's capital. Improved gross margins were achieved through greater milk output per cow, higher concentrate consumption per cow, and higher nitrogen use per hectare. Stocking rate was, however, sufficiently increased so that total variable costs per cow did not increase, despite the higher concentrate costs. By contrast, a study of specialist grassland dairy farms showed that in this situation concentrate costs could be reduced to give improved margins per hectare despite the necessary reduction in stocking rate to allow for the greater reliance on grass. The advantage in gross margin was reflected in improved income and a greater return on capital.

Mr A. Walsh, in his Rex Paterson Memorial Study of thirty-four top grassland dairy farms[1], concluded that 'making full use of grass relies on having faith in its ability to fulfil a more demanding role in dairy herd nutrition. Such faith can only stem from growing grass in sufficient quantity and presenting it to the cow at a satisfactory stage of growth and quality.'

Milk from grass

The MMB Farm Management Services (FMS) fully costs the performance of over 1100 herds in England and Wales. On a sub-sample of 100 specialist dairy herds, a more detailed evaluation has been made of financial performance[2]. In addition, the performance of specialist grassland farms has been analysed and compared with the top 25 per cent of FMS-costed farms, selected on profit per hectare[3]. Together, these reports provide an opportunity to look closely at technical as well as financial performance to answer questions such as: How do top herds compare with the average? How do they generate their extra margins? Do top herds make higher margins per cow *and* per hectare? How much money is made from milk in areas where grass grows well?

In table 1.1 the top quarter of all FMS herds are compared to the average in terms of technical performance and gross margins. Top herds produced more milk and used slightly more concentrates per cow than the average. More significantly, the farmers used more nitrogen and stocked their land more heavily so that, despite spending more money per hectare on fertiliser, forage costs *per cow*

Table 1.1   Top herds compared to average, 1981–82.

	Average	Top herds*
Milk yield		
(litres/cow)	5272	5748
Concentrates		
(tonnes/cow)	1.76	1.82
(kg/litre)	0.33	0.32
Purchased feed costs		
(£/cow)	242	248
Nitrogen		
(kg/ha)	246	299
Forage costs		
(£/cow)	56.5	52.2
Gross margin		
(£/cow)	391	465
(£/ha)	781	1149

* Top 25% selected on gross margin per hectare
Top herds achieved higher milk yields but had similar costs of feed and forage per cow. They used more nitrogen and stocked grass at 2.5 livestock units/ha compared to 2.0/ha on average.

were slightly lower. As a result of this, and the greater income from sales of milk, gross margin per cow was £74 higher and, because the stocking rate was higher, top herds had a gross margin advantage per hectare of £368.

The components of this success are clear. In table 1.2 we can see the percentage contribution of the various technical and financial factors to the extra gross margin per hectare achieved by the top herds. Because feed and forage costs per cow were not very different, these factors contributed little to the extra margins. The overwhelming influence of stocking rate and milk yield per cow in achieving higher margin per hectare is obvious. These two components accounted for 94 per cent of the difference between top herds and the average.

Table 1.2 Percentage contribution to extra gross margin per hectare of top herds.

Higher stocking rate	59
Higher milk yield per cow	35
Higher milk price per litre	3
Lower cost of concentrates and of purchased bulk feeds per cow	−3*
Lower forage costs per cow	2
Other factors	4

* A negative value indicates higher costs for top herds
Higher stocking rate and higher milk yield per cow together accounted for 94% of the extra gross margin per hectare.

Increased milk yield in the top herds was not achieved by feeding more concentrates. Feeding rate was slightly lower at 0.32 kg/litre of milk for top herds, compared to 0.33 kg/litre for the average. Clearly a number of factors were responsible for the higher yield, including the possibility that the top herds were on better grass-growing land. Stockmanship and cow yield potential were also likely to have been above average.

On some farms, especially in areas where grass growth is limited by low rainfall, there is little scope for increasing milk yield and margin per hectare other than by feeding more concentrates to reduce the need for grass. Increased reliance on grass would mean reducing stocking rates. Generally these farms also grow arable crops, and there is motivation to maximise stocking rate to enable a higher area of cash crops to be grown.

In contrast, farmers in areas where grass grows well may have quite different motivations. If the farm is all-grass, other enterprises are likely to be less profitable than dairying; there may also be restrictions to expansion of the dairy herd because of limited labour or capital.

The MMB report[3] of the financial performance of progressive specialist grass farmers producing milk with relatively high reliance on grass, showed that their herds were larger, but milk yields per cow were lower than those in the top 25 per cent of FMS herds. Concentrate use was just over one tonne per cow, and at 0.21 kg/litre was similar to the average (0.23) of the top grassland milk producers in the Rex Paterson Memorial Study. Stocking rate was lower than that of the top FMS herds despite higher use of nitrogen per hectare, because of the greater reliance on grass by the specialist grass herds, see table 1.3.

Looking at the financial performance of the grass farms as shown in table 1.4, the average whole-farm gross margin· was higher than the top FMS herds, but profit per hectare was very similar. The larger herd size (thirty more cows) was associated with higher costs for paid labour, though other overhead charges were very similar. When this difference in labour cost is removed, as it is when management and investment income is calculated, the advantage to the grass farms reappears, largely as a reflection of lower variable costs. This in turn is associated with a higher return on tenant's capital.

Specialist grass herds therefore reduce inputs and sacrifice maximum milk yield, but they produce a level of profit similar to that of top FMS herds. The challenge to the grass farmer is to raise yields economically to increase profit and at the same time maintain the advantage in management and investment income and in return on capital. Another challenge to both groups is to increase stocking rates by economical use of fertiliser and improved grassland management.

Beef from grass

Grass accounts for a greater proportion of the diet of beef cattle than of dairy cows. It is not surprising, therefore, to discover that top MLC-recorded beef farmers achieve higher rates of animal growth from a lower input of concentrates.[4] In other words, they present their animals with an adequate supply of grass of high quality. To achieve this, they apply more nitrogen per hectare. They also stock the grass more intensively so that *per head*, forage costs are contained. That is, they exploit the extra grass they grow. Similar trends emerge from a study of a wide range of systems of produc-

Table 1.3 Specialist grass farms compared to top FMS herds, 1980–81.

	Specialist grass	Top FMS* herds
Cows in herd	134	106
Milk yield (litres/cow)	5308	5643
Concentrates (kg/cow)	1120	1729
(kg/litre)	0.21	0.31
Nitrogen (kg/ha)	302	232
Stocking rate (LSU/ha)†	2.08	2.14

* Top 25% of FMS herds on profit/hectare
† Livestock Units per hectare
Specialist grass herds were larger but had a lower concentrate input and milk
yield per cow; stocking rate was lower despite using more nitrogen per hectare.

Plate 1.1 In areas where grass grows well, margins can be improved by increasing
the proportion of grass in the cow's diet.

Table 1.4   Overheads and profit for specialised grass farms compared to top FMS herds, 1980–81

	Specialist grass	Top FMS* herds
	£ per hectare	
Whole-farm gross margin	750	684

			800	
	1		700	
			600	
Overheads	2		500	
	3		400	
	4		300	
	5			
			200	
Profit			100	
			0	

	Specialist grass	Top FMS herds
Management and investment income	230	164
Return on tenant's capital	20%	13%

1 LABOUR   2 MACHINERY   3 BUILDINGS   4 MONEY   5 OTHERS

* Top 25% of FMS herds on profit per hectare
Specialist grass herds had higher labour costs, but made similar profit per hectare. Return on tenant's capital was higher for the grass herds.

tion. In table 1.5 the data for 18-month beef production are illustrated, and in table 1.6 a summary of suckled calf production is presented.

Top suckler beef producers also achieved greater efficiency of reproduction. Their cows were back in calf 10 days earlier than the average and the calving period was three weeks shorter. They calved 3 per cent more cows and weaned 2 per cent more calves. In addition, the sale weight of weaned calves was 30 kg heavier than the average.

Table 1.5   18-month beef : top herds compared to average, 1981.

	Average	Top third *
Output		
(£/head)	341	356
Daily gain (kg)	0.75	0.80
Concentrates		
(t/head)	0.99	0.8
(£/head)	123	110
Nitrogen		
(kg/ha)	168	192
Forage costs		
(£/head)	40	39
Gross margin		
(£/head)	153	182
(£/ha)	489	672

* Selected on gross margin per hectare
Top herds achieved a higher daily gain and output despite a lower
input of concentrates. They used more nitrogen and stocked their
land at 3.7 cattle/ha compared to 3.2 per hectare on average.

Table 1.6   Top suckler herds compared to average, 1979–80.

	Average	Top third*
Output		
(£/cow)	220	244
(% calves weaned)	93	95
(kg/calf/cow/year)	248	278
Concentrates		
(t/cow+calf)	0.31	0.28
(£/cow+calf)	42	35
Forage costs		
(£/cow)	25	26
Gross margin		
(£/cow)	138	168
(£/ha)	203	277

Top herds achieved greater output through weaning a higher
percentage of heavier calves per cow put to the bull. They did
this from a lower input of concentrates per cow and similar
forage costs. Top herds stocked at 1.65 cows per hectare
compared to 1.47 on average.

In table 1.7 the components of success in 18-month beef and suckler beef production are identified. Improved technical performance, as measured by higher stocking rates, higher sale weight, lower concentrate usage and greater reproductive efficiency accounted for most of the extra gross margin per hectare achieved by the top farmers, with stocking rate the single most important component. Higher stocking rates were sustained by higher levels of nitrogen fertiliser so that nitrogen applied *per head* was similar between average and top herds.

It is often argued that successful beef producers buy well and sell well, especially when overwintering or finishing store cattle. The MLC studies of recorded farmers show clearly that in all cases, good technical performance is the key to financial success. Even in the case of store cattle feeding (see table 1.8), the purchase price per kg and sale price per kg together accounted for no more than a quarter of the extra gross margin achieved by the top third of producers. This is not to say that buying and selling are unimportant; it simply reinforces the view that it is essential to achieve economical weight gains in order to increase output without incurring high costs.

In winter, cattle should have *ad lib* access to high-quality conserved forage, correctly supplemented with concentrates. At pasture, the provision of an adequate daily allowance of high quality herbage

Table 1.7   Percentage contribution to extra gross margin per hectare of top 18-month beef and suckler herds.

	18-month beef	Suckler beef
Technical performance		
Higher stocking rate	42	32
Higher sale weight	10	23
Lower concentrate use per head	22	8
Higher weaning %	—	17
Lower herd replacement costs	—	5
Financial factors		
Higher sale price per kg	6	4
Lower forage costs per head	2	−1
Lower calf price	7	—
Other factors	11	12

Stocking rate was the most important component of success. Technical performance accounted for 74% and 85% of the extra margin of top herds in 18-month and suckler beef, respectively.

Table 1.8 Percentage contribution to extra gross margin per hectare of top third herds of store cattle.

	Lower purchase price per kg	Higher sale price per kg
Winter finishing	11	16
Grass finishing	8	11
Overwintering	11	12

throughout the grazing season is now recognised as a major challenge to the farmer who wishes to make more money from cashing grass with beef cattle. Some farmers have a tendency to deplete grass supply in mid- and late season by cutting too great an area for silage in the early part of the season. This has resulted in reduced performance at grass which has not been rectified in the following winter period.

Flexibility in grassland management is essential, particularly when young beef cattle are reared at pasture, or when store cattle or suckled calves are finished on grass. Reducing risk and making the correct decisions are two key elements in ensuring success in the production of beef from grass. The overall goal must be to improve the predictability of beef cattle growth so that plans laid at the outset for the beef system are realistic and are reflected in the actual level of performance.

Plate 1.2 Top beef producers achieved higher daily gains from their cattle and at the same time had higher than average stocking rates on their grass.

Top beef producers therefore achieved higher daily gains from grass than the average. They used more nitrogen fertiliser and stocked their land more intensively to produce a greater level of output and higher margins per hectare. The challenge for the future is to make further improvements in performance to fully realise the potential for weight gain from grass. In the longer term, gains of at least 1.0 kg/day should be possible from well-stocked grass/clover pastures receiving the correct inputs of fertiliser.

Lamb from grass

Lamb is normally produced from grass with relatively little input of concentrates. The aim is to grow lambs to the optimal weight for slaughter – 18 kg carcase weight in the case of lowland flocks – as rapidly as possible. Early slaughter is favoured to maximise return per kg of lamb carcase.

In contrast to suckler beef, where output is principally measured as the weight of calf sold per cow per year rather than as weaning percentage, in spring lambing flocks the number of lambs reared per ewe is the most important measure of output. Sale returns also depend on the proportion of the lamb crop which is sold for slaughter rather than for a further period of feeding.

Top MLC-recorded sheep flocks sold more lambs per ewe at a higher price, and also sold a higher proportion directly for slaughter than the average (see table 1.9). Returns from wool and from the ewe premium were similar.[5] The higher output was achieved by top flocks at a lower cost per ewe, both for concentrates and for forage.

In contrast to suckler beef where size of herd was similar between top third farmers and the average, top flocks were smaller than the average (see table 1.10). There was no difference, however, in the ratio of ewes to rams which was 38:1 for both the top third and the average.

Stocking rate was overwhelmingly the major component of success, particularly in upland flocks (see table 1.11). The difference in nitrogen fertiliser use per hectare was not very great, and nitrogen applied *per ewe* was similar between top flocks and the average. Possibly, the top flocks were on better land, or had greater reliance on clover-based swards to give enhanced production during the growing season. Alternatively, better management of the grazing areas could have been responsible for improved lamb performance at the higher stocking rates.

Although growing conditions were less favourable for upland

Table 1.9   Top sheep flocks compared to average, 1981.

	Average	Top third
Output		
(£/ewe)	43	49
Lambs reared (%)	143	149
Lambs sold finished (%)	59	67
Concentrates		
(kg/ewe+lamb)	53	50
(£/ewe)	6.6	6.0
Nitrogen		
(kg/ha)	152	170
Forage costs		
(£/ewe)	5.7	4.8
Gross margin		
(£/ewe)	28	35
(£/ha)	351	541

Top flocks reared more lambs per ewe and sold a higher proportion finished. They also received a slightly higher return per kg lamb carcase (£1.86 against £1.78/kg). Higher output was achieved at lower cost per ewe. Top flockmasters stocked their land more heavily than the average (see table 1.12).

Table 1.10   Size of top sheep flocks compared to average.

	Average	Top third
Ewes put to the ram:		
Lowland flocks	460	404
Upland flocks	500	382

flocks than for lowland flocks, the top third of upland flocks had stocking rates higher than the average for lowland flocks (see table 1.12). This was despite the fact that the average level of fertiliser N use by the lowland flocks was 55 kg/hectare more than that used by the top third of upland flockmasters.

Clearly there is tremendous scope for improving grass use in lamb production. Some flockmasters already generate gross margins well in excess of £500 per hectare. In view of the dominant effect of stocking rate on margin per hectare, it would be wise to concentrate on this

Table 1.11   Percentage contribution to extra gross margin per hectare of top flocks.

	Lowland flocks	Upland flocks
Technical performance		
Higher stocking rate	38	68
More lambs reared per ewe	16	9
Lower flock replacement costs	13	6
Lower feed and forage costs	12	5
Financial factors		
Higher sale returns per lamb	17	13
Other factors	4	−1

Stocking rate was by far the most important component of success. Technical performance accounted for 79% and 88% of the extra gross margin of top lowland and upland flocks respectively.

Table 1.12   Increased stocking rate: the key to higher margins in lamb production.

	Average	Top third
Ewes per hectare		
Lowland flocks	13	16
Upland flocks	11	15

single component of success. Improvement here would almost certainly have a greater impact on the enterprise than a change in any other component, especially under upland and hill conditions. The challenge is how to achieve greater stocking rates without at the same time reducing the proportion of lambs sold finished. Improved pasture growth and quality are vital factors in determining whether these objectives can be achieved.

Conclusion: match stocking rate to grass production

One component of success dominates in milk, beef and lamb production — stocking rate. Work at the Grassland Research Institute has shown that with dairy cows wastage of grass is most likely to occur in spring, when grass production exceeds the needs of the animal. Yet there is a reluctance to stock as heavily as is necessary at this time for fear of running out of grass later, when growth is often limited by lack of rainfall.

Plate 1.3 Stocking rate is the single most important component of success in producing lamb from grass. The challenge is to increase stocking rate without at the same time reducing the proportion of lambs sold finished.

The Rex Paterson Memorial Study involved thirty-four top grassland farmers who were already achieving high output of milk from grass. Alf Walsh made a careful study not only of the farms but also of the farmers. He made two important observations, firstly, 'The farmers did not stock their land as heavily as they thought they did, or as heavily as they told me they did!' Secondly, 'They would have been well advised to cut an extra field for silage at the time of the first cut.' If keen producers of milk from grass are under-using spring grass of high value, then the scope for increasing the use of this resource is enormous.

References

1. Walsh, A. (1982) *The Rex Paterson Memorial Study*, British Grassland Society.
2. Amies, S.J. and Craven, J.A. (1982) *Farm Management Services Report No. 33*, MMB.
3. Taylor, K. (1982) *Farm Management Services Report No. 32*, MMB.
4. MLC (1981) *Commercial Beef Production Yearbook*.
5. MLC (1982) *Commercial Sheep Production Yearbook*.

2 Grass Production

In Chapter 1, stocking rate was highlighted as the most important component of success in grassland farming. Top farmers used more nitrogen and stocked their land more heavily so that forage costs per head were no greater than those of the average. Even the top milk producers, however, could have stocked their land more tightly in spring and made more silage.

'Preferred' species

A high yield of grass is the key to achieving higher stocking rates. There should therefore be a close relationship between the level of nitrogen applied to the grass sward, its yield, and the stocking rate of animals per hectare.

But grass swards can deteriorate as they become older. The 'preferred' species of ryegrass, timothy, cocksfoot and clover become replaced in the sward by indigenous species such as meadow grass, bent, Yorkshire fog and other grasses and broad-leaved weeds. In theory a sward with a low proportion of 'preferred' species is less productive, less able to respond to fertiliser and less able to sustain a high stocking rate.

There is support for this theory in the GRI/ADAS Permanent Pasture Group's *Fourth Report*[1] which indicates that the percentage of perennial ryegrass is the only factor correlated with stocking rate. But this may simply show that on the 200 farms in the survey, the fields which had more ryegrass and which were stocked more heavily were those which received more fertiliser, or were on the most fertile soils, or were younger than those which had less ryegrass in the sward.

Recently, ADAS agronomists have found that under conditions of similar soil fertility and fertiliser nitrogen, production is similar between a young ryegrass re-seed and an old but well-managed permanent grass sward with a low proportion of ryegrass (see table 2.1). Although the high ryegrass swards outyielded the low ryegrass ones in spring, this was reversed in mid- and late season. Thus rye-

Table 2.1   A high proportion of ryegrass does not always mean greater
production: comparison of yields in high and low ryegrass swards,
tonnes dry matter per hectare.

	High ryegrass (95% of ground cover)	Low ryegrass
Trial 1	14.7	13.0
Trial 2	14.3	14.5

In both trials, plots were cut four times in 1982 and received 400 kg fertiliser-
N/ha. Despite the low proportion of ryegrass (25% and 5% in trials 1 and 2
respectively), yield was depressed by 12% in trial 1 alone.

grass is most useful when the aim is to generate high yields in spring,
particularly for silage. When the aim is to graze the pasture regularly
during the whole season, then the mid-summer growth of indigenous
species such as *Agrostis* is likely to be of value in maintaining a high
stocking rate.

Trials involving pure stands of different grass species, notably
at Great House EHF and at the West of Scotland College of Agriculture, showed small differences between 'preferred' and indigenous
species at less than 150 kg N/ha, but ryegrass out-yielded all others at
higher rates of N application. However, ryegrass showed a marked
advantage in digestibility, and thus there may be an intake bonus
with ryegrass. For ease of management it helps to have fields of
uniform grass type.

Plant population

The population of grass plants is particularly important in newly
sown leys. If the plants are not there, they cannot produce. Bare
ground is worse than useless — it usually becomes occupied by broad-
leaved weeds thereby restricting the opportunity for grass to increase
in density by tillering.

Look for bare ground in early spring. Grass plants should occupy
more than 90 per cent of the ground area. Large areas of bare ground
signal a patching-up operation or even a complete re-seed. A high
density of grass plants means that the yield response to fertiliser will
be close to that which would be predicted from trials, taking account

of soil type and climate. It also means increased resistance to poaching in wet weather.

An open sward structure and bare ground are encouraged by strip-grazing and the taking of heavy cuts for conservation. Intermittent, rather than continuous defoliation leads to tiller extension rather than tiller initiation, thus tiller population can decline relative to a set-stocked sward. Strip-grazing is therefore a recipe for sward deterioration. Prevent it by continuous close grazing, or by alternate cutting and grazing.

Temperature and grass growth in spring

Early bite is at its most valuable when there is no more silage or hay left; it can also provide a boost to performance, particularly if the animals have been rationed, or given an unbalanced diet in late winter.

Some farmers go to great lengths to secure early grass, for example by sowing a crop of winter rye. Others record the temperature each day to assist in the timing of early dressings of nitrogen.

Accumulated temperature should provide a useful guide as to how quickly the soil is warming up in late winter and early spring before the grass starts growing. Application of nitrogen to boost the growth of grass, once it starts, should therefore bear some relation to changes in temperature. In this way a balance can be struck between applying nitrogen too early, risking loss by denitrification (gaseous loss of nitrogen from the soil) or leaching (as a result of excessive rainfall), and applying it too late with loss of grass yield. The aim is to have sufficient nitrogen available when growth commences, to maximise the yield for early grazing and first-cut silage.

Timing nitrogen for spring grass

The T-sum 200° system

The T-sum system is based on trials carried out over many years in Holland and evaluated, since 1979, in the UK. Daily air temperature is averaged and accumulated from 1 January. Negative *averages* (not negative temperatures) are ignored (see box). The first dressing of nitrogen is applied when the T-sum reaches 200°C, provided the land is dry enough.

How to calculate T-sum

1. Buy a maximum and minimum thermometer, put in on an exposed wall near the house.
2. Record the *average* air temperature in degrees Centigrade from 1 January at the same time each day.
3. Add up the daily averages, ignoring negative averages.

Other systems

The T-sum 200° is not the only system. Two others both based on soil rather than on air temperature have been evaluated by the Scottish Colleges. It was felt that soil temperature was perhaps a more logical measurement to take, since plants grow in soil and not in air. A further difference between the two Scottish methods and the T-sum 200° method is that recording starts on 1 February, not 1 January.

These two systems are called T-value 100°, in which positive soil temperatures recorded at 100 mm depth are accumulated at 0900 h each day from 1 February, and T-soil 150° in which soil temperatures at 300 mm depth are accumulated. Since 1979 these systems have been compared with T-sum 200° in trials at the West of Scotland College of Agriculture.

Measured over ten trials, T-sum 239°, T-value 89° and T-soil 117° gave the best yields. Whatever system is used, there is approximately one week for each 25° change in accumulated temperature, so warnings of T-sum 200°, T-value 50° and T-soil 75° would assist in the timely application of fertiliser.

All three systems can therefore be used to predict the best time to apply nitrogen. There was a considerable range of accumulated temperature within each system at which the best yield was obtained. Thus there was a period of two to three weeks around T-sum 200° at which 90 per cent or more of the maximum yield was obtained.

Trials carried out by ADAS, involving fifty-two sites over three years, have confirmed that there is a period of two to three weeks *after* T-sum 200° before there is a marked decline in yield due to application being too late. T-sum 200° almost invariably comes within the period when yield is 90 per cent or more of the maximum.

All the authorities agree there is flexibility over the accumulated T-sum at which nitrogen should be applied. T-sum 200° is not the *only* correct time to apply nitrogen, but it *is* the only time that is rarely wrong.

Effect of site on T-sum

The two main influences on the rate of accumulation of temperature are the altitude and latitude of the farm. The map in fig. 2.1 shows the average date when the T-sum 200° was reached, using records over a twenty-year period kept by the Meteorological Office. Wick and Kinloss, just above sea level on the north-eastern coast of Scotland reached T-sum 200° on 22 February, whilst High Mowthorpe in Yorkshire, at 200 metres above sea level was ten days later. By contrast, Plymouth, also at sea level but on the south coast, was twenty days earlier.

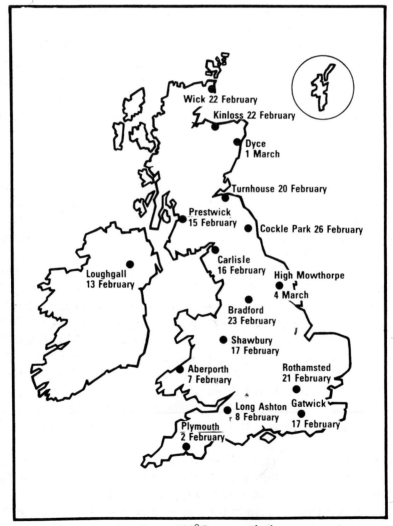

Fig. 2.1 Average date when T-sum 200°C was reached over a twenty-year period.

It might be argued that as there is a two to three week period around T-sum 200° during which nitrogen may be applied, there is little point in recording daily average temperatures. A 'target' calendar date of T-sum 200°, supplied by the local meteorological office, might be good enough.

T-sum and total annual yield

There is no evidence that the application of fertiliser nitrogen at different T-sums affects total annual yield. But since most trials have concentrated on the first cut at a grazing stage of growth, or on the first cut for silage, more information is needed, particularly in relation to silage cuts. Applying N at T-sum 200° rather than at 300° or 400° might allow the first grazing or the first cut of silage to be taken slightly earlier than would otherwise be the case. The yield would then be similar although the date of cutting would be different. In areas where the supply of water for grass growth in mid-season is low, an earlier first cut might allow the crop to regrow to a greater extent before the onset of drought, to give a higher yield at the second cut.

T-sum for first cut silage

Most farmers would like to make more silage, so that the requirements of the animals can be met, not only through the winter period but also into late spring if the weather is wet. Yet there is little information on the effect of different times of application of nitrogen on the yield of first-cut silage.

Trials by UKF Fertilisers Ltd in 1981, 1982 and 1983 showed a small but significant yield advantage to a split dressing of 90 kg N per hectare at T-sum 200° followed by 60 kg N per hectare in early April, compared to a single dressing of 150 kg N in early April. Plots were harvested at an average D-value of 68 (see fig. 2.2). A double application is more costly, but it introduces flexibility into the management of grass at a critical time of the year. Many farmers would be well-advised to tighten up on grazing stocking rate in the early part of the season and to have a buffer area which could be either grazed if grass growth is slow, or cut for silage if grass growth is rapid.

Application of fertiliser at T-sum 200° on the silage area as well as on the grazing land should therefore make more grass available for grazing if growth is slow. Conversely, because the whole grass area has received nitrogen early in the season, a greater proportion can be set aside for silage if grass growth is rapid. If the buffer grazing

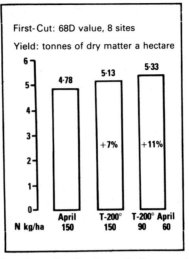

Fig. 2.2 T-sum 200° for silage. Application of nitrogen at T-sum 200° gave a 7% improvement in yield whilst a split dressing gave an 11% improvement over a later, single dressing in April.

system is adopted, it too should receive nitrogen at T-sum 200° since it will benefit whether it is grazed or cut.

If grass growth is very rapid in March giving an abundance of herbage in early April, the second application of nitrogen may not be necessary.

Losses of nitrogen from grass in spring

The main losses of nitrogen in early spring are due to leaching and denitrification. Leaching occurs as a result of wet weather, and denitrification is generally enhanced by above-average temperatures in early spring. Trials at the permanent grassland station of the Grassland Research Institute at North Wyke in Devon in 1982, showed that first-cut yields were low from plots which received fertiliser (70 kg N per hectare) when the soil was waterlogged in March between T-sum 300° and 390°. Substantial quantities of N were probably lost during this period of high rainfall due to denitrification, leaching and surface run-off. Application of nitrogen to the same poorly-drained clay soil either earlier, between T-sum 200° and 300°, or later, between T-sum 400° and 600°, gave higher yields.

The principal chemical entity involved in N losses by either denitrification or leaching is nitrate, which comprises half the N when ammonium nitrate is the fertiliser. The conditions conducive to denitrification are:

(a) a warming soil, between 5°C and 8°C;
(b) restricted aeration, therefore a high water content in the soil;
(c) a high nitrate content of more than 5 to 10 kg/hectare in the top 20 cm of the soil.

Under these conditions the soil micro-organisms strip the oxygen from the nitrate to leave nitrogen gas or nitrous oxide (laughing gas), both of which are then lost from the soil to the atmosphere. Ammonium ions, on the other hand, are relatively immobile. Being positively charged cations they are absorbed on to negatively charged clay particles. Thus they are not leached out by rain, nor are they lost by denitrification.

The GRI team[2] studied the losses from 70 kg N per hectare, applied on either 8 February 1982 (close to T-sum 200°) or in mid-March 1982, as either ammonium nitrate or ammonium sulphate. The results are shown in table 2.2. Two features are apparent: first, losses were lower when ammonium sulphate was used compared to ammonium nitrate, and yields of dry matter were some 26 per cent higher at the first cut. Second, the loss due to denitrification was greater when ammonium nitrate was applied in mid-March rather than in February at T-sum 200°.

The main problem with ammonium sulphate is that it acidifies the soil so that three times as much lime is required for every 100 kg nitrogen as compared to using ammonium nitrate. An alternative may therefore be to inject anhydrous ammonia into the sward, or to use urea.

Table 2.2   Loss due to denitrification and yield of first-cut grass: different fertilisers compared.

Date of application	Denitrification loss kg N/hectare	Dry matter yield* t/hectare
8 February		
Ammonium nitrate	3.1	1.12
Ammonium sulphate	0.6	1.41
15 March		
Ammonium nitrate	10.5	0.95
Ammonium sulphate	0.5	1.21

* Between date of application and 10 May 1982

In 1983 urea was 30 per cent cheaper than ammonium nitrate, per unit of nitrogen. Very similar yields were obtained in trials with urea compared to the equivalent amount of nitrogen from ammonium nitrate. However, the size of urea prills is variable and this may pose problems in achieving accurate application. Loss of N due to denitrification was low, but the high pH conditions surrounding the urea prills was reflected in volatilisation of ammonia and loss as ammonia gas. Wet weather following application of urea would help to prevent this loss since the ammonia would then be washed into the soil as ammonium hydroxide, and the ammonium ions would be absorbed on to the soil particles. Furthermore, when root temperature is between 5°C and 12°C, as it is in March and early April, ammonium-N rather than nitrate-N is the preferred form of uptake of N by the grass plant.

Finally, at whatever time fertiliser is applied, care should be taken to reduce damage to fields by machinery. Waterlogged soils should be avoided altogether, except during periods of frosty weather. Low ground-pressure vehicles are useful for minimising damage during spreading and if available they should be the preferred type of equipment during late winter and early spring when soils are at field capacity.

Predicting silage time from soil temperature

Soil temperature may help in predicting heading date, and in planning when to cut for silage. Analysis of twenty years' data revealed a close correlation between the average soil temperature at 30 cm depth for the month of March and heading date in S24 perennial ryegrass[3]. The results are summarised in table 2.3, together with estimated heading dates for later-flowering ryegrass varieties.

Some worthwhile research would be to find out if the T-value 100° method (which involves measuring soil temperature at 10 cm, not at 30 cm) could be extended to give an early prediction of heading date. If so, this would help greatly in planning in early April for silage of specified D-value in May.

Heading date in ryegrass corresponds to a D-value of about 67, or a metabolisable energy (ME) value of 10.5 MJ/kg DM. This is the target average quality for a three-cut silage-making regime. But to achieve this quality, it is necessary to start cutting earlier than the heading date, depending on how many days are required to harvest the total area.

Table 2.3   Prediction of heading date from soil temperature.

	S24 ryegrass	Later-flowering ryegrass
Average 30cm soil temperature in March (C)	Heading date	
10	April 27	May 9
8	May 5	May 17
6	May 12	May 24
4	May 20	June 1

Heading date (50% ear emergence), occurs at 68 D-value for S24 and at 65 D-value for later-flowering varieties (e.g. Melle, Meltra), which are estimated to flower twelve days later than S24.

Recommended levels of fertiliser for grass

The amount of nitrogen (N) applied to grass is the most important factor affecting grass growth, but the response to N can only be realised if phosphorus (P) and potassium (K) are also in adequate supply. The response also depends on the potential of the field for grass growth, i.e. the type of soil and the amount of summer rainfall.

The current recommendation for nitrogen for grazed swards is 2.0 to 2.5 kg/ha per day of the growing season. In practice this means a target, depending on the site, of 300 to 450 kg N/ha over the whole season. In contrast, the average level of application on dairy farms is only 170 kg N/ha, and considerably less than this on beef and sheep farms.

Up to the target level, the response to N is about 20 kg dry matter per kg N; above the target level, the response falls to less than half this rate and is barely economic.

To avoid hypomagnesaemia (grass staggers), compound fertiliser or slurry should not be applied to grazing areas in spring.

Assuming a phosphate and potash status of 1, the recommended levels of N, P_2O_5 and K_2O for grazed and conserved grass are as shown in fig. 2.3.[4] It is advisable to check P and K status by analysing the soil or herbage in late May and adjusting subsequent applications if necessary.

Slurry should be applied to grazing land immediately after a close grazing. Because K is the limiting factor to its use, the upper limit of application of undiluted slurry on grazed pastures is 7 m^3/ha (7,000 litres/ha). Slurry can contribute to the nutrient requirements of conserved grass to a greater extent than grazed grass. Because of its relatively high content of available K, it can be used to supply all the K that is needed by the crop. At the levels shown in fig. 2.3, slurry will also supply about 25 per cent of the requirement for N and about 40 per cent of the requirement for P for each cut. If the recommended rates of application are followed, all the slurry produced from a herd of dairy cows can be used effectively on the conservation and grazing land.

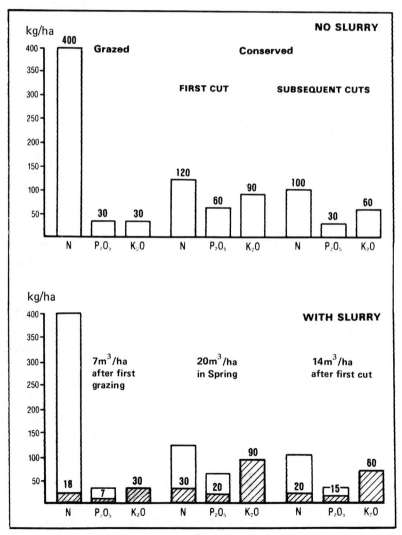

Fig. 2.3 Recommended levels of fertiliser for grass (kg/ha).

Plate 2.1 The target for fertiliser nitrogen is 300 to 400 kg/ha over the whole season. Slurry can supply up to 25% of the requirement for N, 40% of the phosphate and 100% of the potash required for each cut.

Nitrogen for grass/clover swards

It is probably true to say that of the 'preferred' species of plants in grassland, clover is the most elusive. Not surprisingly, therefore, a considerable research programme is currently underway in the UK to find cost-effective ways of establishing and maintaining clover in grass swards.

Perhaps one reason why the use of fertiliser-N on grass is so much lower than current recommendations is that many grass farmers already attempt to maintain clover in their swards in the presence of limited fertiliser-N. The development of longer-petioled varieties of clover which can tolerate higher levels of N is, however, a welcome advance and one which might herald the introduction of even more flexible varieties which will tolerate a wider range of environments and managements.

Trials over three years at twenty-one sites, in which grass alone was compared to grass/white clover swards showed the classic responses to fertiliser-N (see fig. 2.4). The white clover/grass swards yielded as much as the pure grass swards receiving 200 kg/ha of fertiliser-N and out-yielded the pure grass sward at all rates of N up to 400 kg/ha.[5]

An interesting trial by UKF Fertilisers Ltd[6] shows the scope for achieving high yields from a combination of fertiliser-N and clover. Even with 150 kg/ha of fertiliser-N, a vigorous clover content was maintained in the sward with both Huia and Blanca varieties over the

Plate 2.2 White clover grows more slowly than grass in spring; very high dressings of fertiliser-N should be avoided in early spring if the potential of clover is to be realised in mid-season.

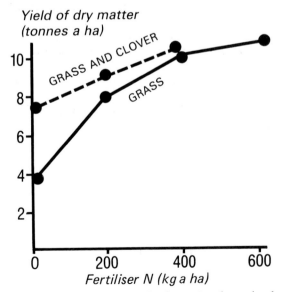

Fig. 2.4 Clover boosts grass production. In the national grassland manuring trial, yields of grass/clover plots cut monthly were higher than for grass alone up to 400 kg/ha of fertiliser N.

Table 2.4   Clover can make a large contribution to yield in the presence of fertiliser N in good grass-growing conditions.

	Grass (S23 ryegrass)	Grass and clover
Yield		
(tonnes dry matter/ha)	7.4	11.8
Clover in sward (%)		
Huia		34
Blanca		48

Yield was 60% higher for grass plus clover than for grass alone.

four years of the trial (see table 2.4).

White clover grows more slowly than grass in spring, but more vigorously in mid-season. It is advisable therefore to avoid very high dressings of fertiliser-N in early spring if the potential of the clover is to be fully realised in mid-season.

References

1. Joint Permanent Pasture Group (1982) *Fourth Report*, GRI/ADAS.
2. Ryden, J.C. *et al.* (1982) *Grassland Research Institute Annual Report*, GRI, 26–28.
3. Roy, M. (1972) *Journal of the British Grassland Society* **27**, 231.
4. ADAS (1982) *Profitable Utilisation of Livestock Manures*, Booklet 2081, HMSO.
5. Morrison, J. (1981) *Proceedings of the Winter Meeting of the British Grassland Society.*
6. Mackenzie, G.H. and Daly, M. (1981) *Proceedings of the Winter Meeting of the British Grassland Society.*

3 Utilised Metabolisable Energy

Utilised metabolisable energy (UME) is the amount of metabolisable energy (ME) grown as grass which is actually eaten by the animal. It is normally expressed as gigajoules (GJ) per hectare (1 GJ = 1000 megajoules, MJ). It is not possible to measure grass consumption directly, so UME is derived from information about the ME required by the animal and the ME supplied as concentrates.

The simplest way of calculating UME per hectare for a dairy herd is shown below:

$$\text{(a) MR required per cow} = 25 + \left[\frac{\text{milk yield (litres/cow)} \times 5.3}{1000} \right]$$

$$\text{(b) } \textit{less } \text{ME from concentrates} = \frac{\text{concentrates (kg/cow)} \times 11}{1000}$$

(c) *multiplied* by stocking rate (cows/hectare).

The amount of UME per hectare is a measure of the useful output from grass. It is therefore a more relevant measure of productivity than the yield of grass alone. Because grass is grown primarily for feed for the ruminant, measurement of its use is a more sensible yardstick of grassland output than is grass production.

High output of UME depends on:

(a) the right plants being present in the sward;
(b) good growing conditions for such plants; and
(c) a high proportion of the grass produced being eaten.

Effect of site on grass growth

Recognising the importance of having the correct species and density of grass plants in the sward, the production of ME depends on the growing conditions which are inherent features of the site such as soil type and rainfall during the growing season. Nearly all of the differences in growth due to water shortage occur in mid- and late summer and autumn.

The classification of site classes is shown in table 3.1. Growing conditions vary from 1 (very good) to 5 (poor).[1] The response in grass production to nitrogen is higher under good growing conditions than under poor conditions, and it is possible to set optimal levels for nitrogen which are economical in relation to site class (see fig. 3.1). Levels range from 300 kg/ha for poor growing conditions to 450 kg/ha for very good growing conditions. The pattern of application can be varied to enhance either spring growth by applying relatively high dressings of 100 kg/ha early in the season, or mid-season production by restricting early season applications and applying relatively more nitrogen after the first cut of silage.

Targets for UME output

The optimal levels of nitrogen may be anticipated to give yields of ME which can then be translated into targets for UME output (see

Table 3.1 Classification of site classes; most grassland areas are in above-average growing conditions.

Site class depends on rainfall and soil type

| Soil texture | Rainfall April to September | | |
	More than 400 mm	300 to 400 mm	Less than 300 mm
		Site Class	
Clay loams and heavy soils	1	2	3
Loams, medium textured soils and deeper soils over chalk	2	3	4
Shallow soils over chalk or rock, gravelly and coarse sandy soils	3	4	5

(Add 1 for northern areas, e.g. Scotland, and for sites above 300 m elevation).

Growing conditions depend on site class

| | Site Class | | | | |
	1	2	3	4	5
Growing conditions	very good	good	average	fair	poor

The range of conditions covers the whole country.

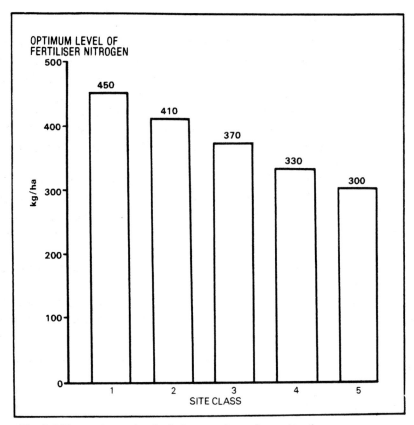

Fig. 3.1 The optimum level of nitrogen depends on site class.

fig. 3.2). The values are based on the results of national grassland manuring trials and on the rainfall expected in seven years out of ten.

High output of UME can be achieved by using a high level of fertiliser nitrogen, or by using a high proportion of the ME which is grown, or both. Thus a UME of 80 GJ/ha may be the result of in-efficient use of a high yield of ME in site classes 1 and 2, or very efficient use of a relatively lower yield at site classes 4 and 5.

The targets for UME shown in fig. 3.2 assume a level of efficiency of utilisation of 85 per cent. High efficiencies are achieved by:

(a) close grazing;
(b) frequent silage cuts;
(c) adequate lime, phosphate and potash;
(d) healthy swards.

Results in practice

There is bound to be a very wide range in UME output on farms. This range reflects differences in grass growing conditions, level of fertiliser, stocking rate, level of concentrate feeding, cow potential and other factors. Nevertheless, there is a trend towards higher levels of UME output from grass, and this is shown clearly in the UME values calculated from the results of the dairy farms costed by BOCM Silcock.[2] Over the period 1966 to 1980, UME output increased from 50 GJ/ha to 70 GJ/ha.

The BOCM sample of herds were not only above average for milk yield, but they also had higher stocking rates than those of other surveys. Consequently, the levels of UME output were relatively high, though only a little higher than those derived from comparable surveys done by ICI and the MMB.

A realistic current average UME output on UK dairy farms is probably about 65 GJ/ha, with a probable range of 30 to 140 GJ/ha. By contrast the *average* UME output on the thirty-four top grassland farms in the Rex Paterson Memorial Study[3] was 104 GJ/hectare (see

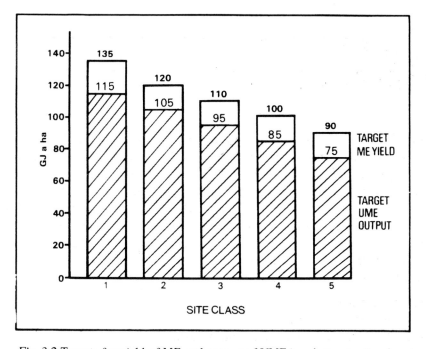

Fig. 3.2 Targets for yield of ME and output of UME in relation to site class.

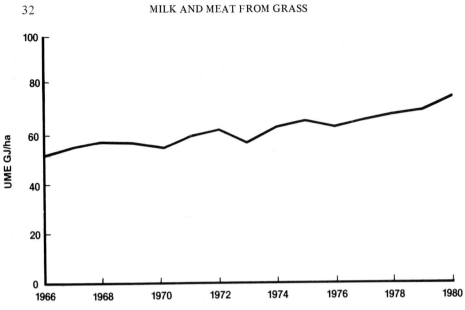

Fig. 3.3 Output of UME/ha is increasing on UK dairy farms.

table 3.2). The range of systems covered by the study was enormous but they all had one thing in common — they were highly efficient in the production of milk from grass.

Just how efficient they were is shown in fig. 3.4 in which the levels of UME output are shown in relation to stocking rate and fertiliser nitrogen. Even those farmers who had stocking rates close to the national average of 1.8 cows per hectare used considerably above-average levels of N (260 kg compared to 170 kg N/ha). At the highest levels of stocking — there were three herds with stocking

Table 3.2   Top grassland herds compared with average recorded herds.

	Average*	Top grassland herds†
UME output (GJ/ha)	65	105
Milk output (Litres/cow)	5262	5946
Stocking rate (cows/ha)	2.0	2.4

* Milk Marketing Board average for FMS herds 1980/81.
† Rex Paterson Memorial Study 1980/81.
Top grassland herds averaged 105 GJ UME/ha. They achieved this by a combination of high milk yield, high stocking rate with economical use of nitrogen (345 kg/ha) and concentrates (1.4 tonnes/cow).

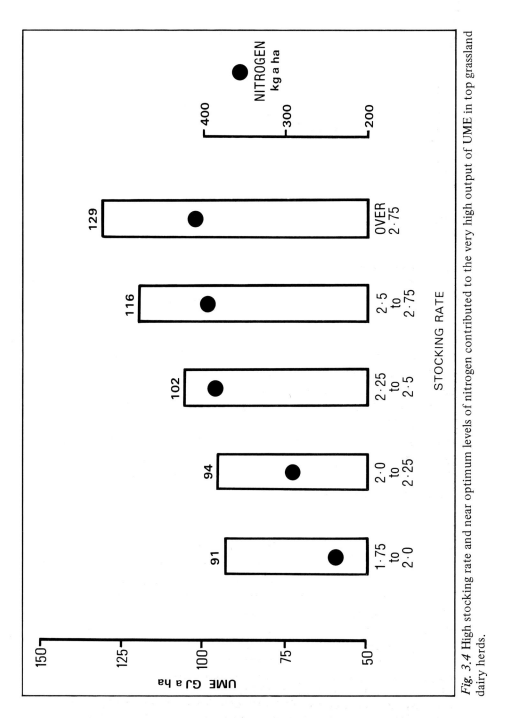

Fig. 3.4 High stocking rate and near optimum levels of nitrogen contributed to the very high output of UME in top grassland dairy herds.

rates above 2.75 cows per hectare — UME output was twice the national average, which gives an indication of the potential in areas where growing conditions for grass are very good.

UME is an important guide to profitability because a unit of ME from grass and forage costs less than half a unit of ME from concentrates. Results from MMB-costed farms[4] shows a close relationship between UME and gross margin per hectare. Thus 10 GJ extra UME per hectare is equivalent to an increase in gross margin of £140 per hectare (see fig. 3.5).

The UME system is a way of setting targets and diagnosing problems on any farm, regardless of its location and its current use of fertiliser and concentrates. It can be a most effective aid to increasing

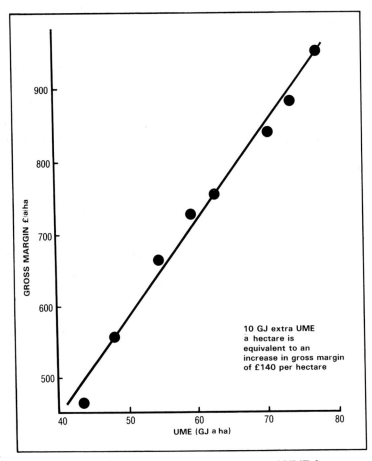

Fig. 3.5 Gross margin/ha is closely related to output of UME from grass and forage.

Plate 3.1 Irrigation of grass may be an attractive way of boosting UME output and margins. A unit of ME from grass costs less than half a unit of ME from concentrates.

profitability through better use of grass. The farms in the Rex Paterson study which had levels of UME of around 130 GJ/ha were making margins over feed and forage costs of £1500 per hectare — more than twice the national average. These herds also used the highest levels of forage ME per cow and at the same time achieved milk yields in excess of 6500 litres per cow.

References

1. Young, J.W.O. (1982) *Farm Advisory Note No. 23*, ICI.
2. BOCM Silcock Ltd (1981) *Dairy Costings*.
3. Walsh, A. (1982) *The Rex Paterson Memorial Study*, British Grassland Society.
4. Amies, S.J. and Craven, J.A. (1982) *Farm Management Services Report No. 33*, MMB.

4 Grazing

Grazing is the one area of grassland management still open to major improvements both in output and in efficiency of use. A measure of the difficulty in achieving such improvements is shown in the current trend away from grazing, as farmers attempt to reduce risk and uncertainty by making more silage and housing their cattle earlier in the autumn. Stocking rate is increased, not by grazing more tightly to reduce losses at pasture, but by increasing the proportion of grass which is cut.

Grass is grazed because it is a low-cost way of providing high-quality feed to animals. Grazed grass is estimated to be half the cost, per MJ of metabolisable energy (ME), of conserved forage and a quarter the cost of concentrates. The challenge is therefore to make the most effective use of grazed pasture, to maximise the output of utilised metabolisable energy (UME), and to reduce the wastage of ungrazed grass to as low a level as possible.

Continuous or rotational grazing?

The choice of grazing system depends mainly on the layout of the farm fields, access tracks, fencing and water supply. Milk production trials comparing continuous (i.e. set stocking) with rotational (paddock) grazing have not shown a consistent advantage to either system (see fig. 4.1).[1] A rotational system which allows greater control of daily herbage allowance is better suited to cows with high milk yield potential, such as spring calvers.

The advantages of rotational grazing are:

(a) greater flexibility in matching the supply of grass to the requirements of the animals;
(b) it is better suited to spring calving cows;
(c) easy matching to small fields;
(d) less time needed for moving animals;
(e) greater suitability for site classes 4 and 5 (fair and poor growing conditions).

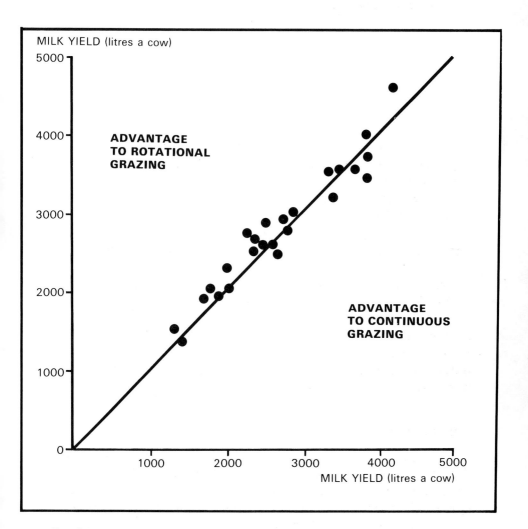

Fig. 4.1 There is little difference between continuous and rotational grazing in production of milk per cow.

The advantages of continuous grazing are:

(a) denser swards with greater resistance to poaching;
(b) swards likely to be more persistent;
(c) less fencing needs;
(d) less time needed for field operations;
(e) greater suitability for site classes 1, 2 and 3 (good grass growing conditions).

There has, however, been a marked move away from elaborate rotational grazing systems in recent years. This is perhaps linked to the fact that the majority of dairy cows calve in the autumn, and are therefore in mid-lactation at the time of turnout.

With beef and sheep stocked at average rates, many farmers have found little difference in individual performance or in output per hectare as a result of moving to a simpler continuous grazing system.

Reducing risk and uncertainty in grazing

Whatever grazing system is adopted, the elements of risk and uncertainty tend to result in pastures being undergrazed in spring to avoid overgrazing and reduced animal performance in mid- and late season.

Various measures can be adopted to reduce the element of risk during the grazing season. They include integrating cutting with grazing, buffer grazing and buffer feeding of supplements to grazed pasture.

Integrating cutting with grazing

The timing and the frequency of cuts for silage both have an influence on the regrowth of aftermaths and hence on the availability of grazing areas in mid- and late season. This effect is greater under poor grass growing conditions than in areas where grass grows well (see table 4.1).

Infrequent cutting of grass for silage leads to greater wastage of grass in grazed areas, particularly under poor grass growing conditions, than with frequent cutting because a fixed area of land is closed

Table 4.1 Reduce wastage of grass in areas set aside for grazing by more frequent cutting, integrated with grazing.

	Four cuts	Three cuts	Two cuts
Silage ME (MJ/kg DM)	10.9	10.7	10.2
Silage D-value	68	67	64
Wastage of grass DM in grazing areas (%)			
Site class 1 (very good)	26	26	27
Site class 3 (average)	22	27	26
Site class 5 (poor)	22	30	37

up for a longer period of time, and more land has to be allocated for grazing in early season. Also a relatively large regrowth area is released for grazing at a time when production of grass is likely to exceed the requirements of the animals.

Buffer grazing

At the other extreme, closing up land for silage increases the risk of running short of grass later in the season. A study by the MLC of farmers operating the 18-month beef system showed that an important factor in reducing liveweight gains at pasture was poor performance in the second half of the season, attributed to a restriction in the supply of aftermath herbage. Farmers were closing up too great an area of land in an attempt to make more silage, regrowth herbage was inadequate to meet requirements, and animal performance over the latter half of the grazing season suffered as a result.

Buffer grazing has been developed at the Edinburgh School of Agriculture[2] as a way of increasing the area of regrowth available in mid-season to set-stocked beef cattle, thereby reducing the risk of running short of grass at this time. A proportion of the grazing area is reserved behind a movable electric fence, to be conserved if possible or grazed in thirds if early grass growth is too slow. For example, in 1981 grass growth in early season was sufficient to allow 4.7 tonnes of DM per hectare of buffer area to be ensiled. The regrown buffer area was then grazed for four weeks before the cattle were turned on to the main silage aftermaths. Cattle turned out at the same initial stocking rate without the buffer gave lower rates of growth and lower overall gains per head and per hectare (see table 4.2). The procedure is shown diagrammatically in fig. 4.2.

Table 4.2 How buffer grazing improves beef cattle gains at pasture.

| | Turnout to aftermath grazing* (kg/day) | Liveweight gains | | |
		Aftermath to housing (kg/day)	Overall (kg/day)	(kg/ha)
Buffered	0.75	0.82	0.77	1068†
Unbuffered	0.66	0.75	0.70	980

* Initial stocking rate was the same for each group, 3200 kg liveweight/hectare
† Plus 4.74 tonnes of silage DM per hectare

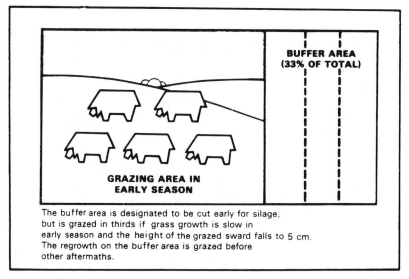

Fig. 4.2 Edinburgh buffer grazing system.

Buffer feeding

As the season progresses, acceptable grass becomes increasingly difficult to find as areas of rejected herbage accumulate. The rejected herbage results from previous fouling and treading. In addition, the rate of grass growth is slower than in spring, and in periods of hot, dry weather, growth can virtually cease completely due to lack of available water.

Cows rarely graze for more than nine hours per day. The consequence of this is that grass intake declines during the season. As grass availability is reduced, the cows cannot fully compensate by increasing their grazing time.

Typical results from trials at the Crichton Royal Farm of the West of Scotland Agricultural College[3] are shown in table 4.3. The average intakes will vary according to grass availability, and individual intake will vary according to cow milk yield, but the general trend for intake to decline towards the end of the season has been observed at other research centres. It reflects the inability of the cow to increase the time spent grazing per day to compensate for the reduced supply of acceptable grass. This decline in intake is not confined to dairy cows. Calves, beef cattle and sheep are also likely to face periods in mid- and late season when the availability of 'good' grass is low.

Recent trials have shown the value of a buffer feed when grass is in short supply. Its value is as a tactical feed which, although it may

Plate 4.1 Buffer grazing: a proportion of the grazing area is reserved behind a movable fence and grazed if grass growth is poor, or cut for silage if grass growth is good.

be on offer every day, is only likely to be eaten when the cow finds her supply of 'good' grass for grazing is reduced. Another important feature of the buffer feed is that it must not be eaten in preference to grass, otherwise it will simply substitute for it and lead to in-

Table 4.3   How intake declines as the grazing season progresses.

	Early Season	Mid-Season	Late Season
Time spent grazing (h)	8	9	9
Grass intake (kg dry matter/day)	16	14	11

creased wastage of grass in the field. Hay or silage are therefore
appropriate for use as buffer feeds.

Concentrates are often given to dairy cows in increasing quantities
as the grazing season progresses, regardless of how much grass is
available for grazing or its quality. This is because autumn grass is
usually considered to be of inferior feed value to spring grass. Studies
at the Grassland Research Institute point to a higher degradability of
nitrogen in the rumen for mid- and late-season grass than for spring
grass. A source of undegraded protein ought, therefore, to be the
most appropriate type of supplement for mid- and late-season grass.

Apart from the problem of degradable nitrogen, the evidence from
grazing trials is that the energy content of the grass which is selected
by the animal remains high throughout mid-summer and autumn. It
is the *quantity* of grass eaten which declines. Dr Leaver has proposed
a simple guide to concentrate supplementation for grazing dairy cows
which takes account of the seasonal decrease in grass consumption,
and the variability which can occur in grazing conditions within the
season. The guide, shown in table 4.4, also incorporates a forage
buffer feed once daily when grazing conditions are judged to be
poor.

Table 4.4   Recommended feeding rates of concentrates to grazing dairy cows.

Grazing conditions	Early Season	Mid-Season	Late Season
	(kg concentrate/kg/milk)		
Poor*	0.2	0.3	0.4
Moderate	0.1	0.2	0.3
Good	0	0.1	0.2

* Provide in addition a buffer feed once daily.

This system of concentrate feeding expects the high-yielding cow
to take more from grass than the low-yielder. High-yielders do, in
fact, eat more grass than low-yielders, but low-yielders are likely to
be producing milk with a higher solids content, and are also gaining
in weight. They are also likely to have a pregnancy requirement for
energy. The net effect is that the difference between high- and low-
yielders in total energy requirement is not nearly as great as might
appear at first sight. Thus the level of concentrate allowance reflects
differences in probable grass intake rather than different stages of
lactation.

Maintaining intake

The decline in intake as the season progresses can be minimised by ensuring that as far as possible the supply of 'good' grass is adequate. This means changing fields regularly and having dense swards of the correct height on offer to the animals.

The plan is to ensure that grass height in grazed areas is neither too short nor too long. A simple guide is the 'wellie test', shown in fig. 4.3. The recommended heights of grass refer to grazed, not rejected areas. With rotational grazing the target height refers to grass post-grazing. In the case of continuous grazing the daily grass height should be maintained at wellington-boot height (6 to 8 cm) at all times. If grass height falls below these targets, intake is likely to be depressed and production will decline. If grass height is above the targets in fig. 4.3 grass wastage will be increased.

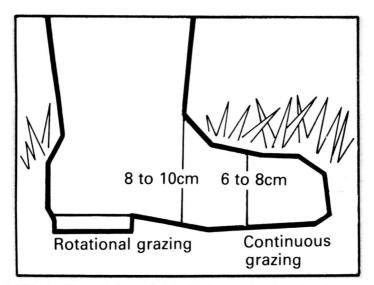

8 to 10cm   6 to 8cm

Rotational grazing   Continuous grazing

Fig. 4.3 The 'wellie test' of grass height. Under rotational grazing the target post-grazing grass height should be 8 to 10 cm in grazed areas. With continuous grazing the height should be 6 to 8 cm or level with the toe of the wellington.

Target stocking rates

The optimum level of fertiliser nitrogen for each site class (see chapter 3) determines the target stocking rate for the early, mid- and late periods of the six-month grazing season. The targets for dairy cows and beef cattle are shown in fig. 4.4 for good (site class 2) and poor (site class 5) grass growing conditions. In each case the

Plate 4.2 Under continuous grazing, daily grass height should be maintained at 6 to 8 cm.

target stocking weight at turnout to grass is 2500 kg per hectare for good, and 2000 kg per hectare for poor grass growing conditions.

The stocking rates refer to an integrated cutting and grazing regime where aftermaths are grazed in mid- and late season. Three cuts are taken during the season to give silage of 10.7 MJ ME/kg DM (67 D-value), in mid-May, end of June and mid-August.

As the season progresses, overall stocking rate is reduced to maintain as far as possible, an adequate supply of acceptable herbage for grazing. Thus the number (and weight) of cattle is halved between early and late season.

Targets for beef cattle growth at grass

Target daily growth rates at grass are between 0.7 kg and 1.0 kg, depending on breed and age of animal. Store cattle finished at grass should reach 1.0 kg/day growth rate. On the other hand, spring-born calves in their first year at grass would only be expected to grow at 0.7 kg/day. Six-month-old Friesian calves should achieve 0.8 kg/day whilst Hereford x Friesians should grow at 0.9 kg/day. These targets are normally not difficult to achieve in the first three months of the grazing season, but thereafter they often begin to look out of reach.

The important point to remember is that gains not made at grass are unlikely to be recovered in the subsequent winter period. As a

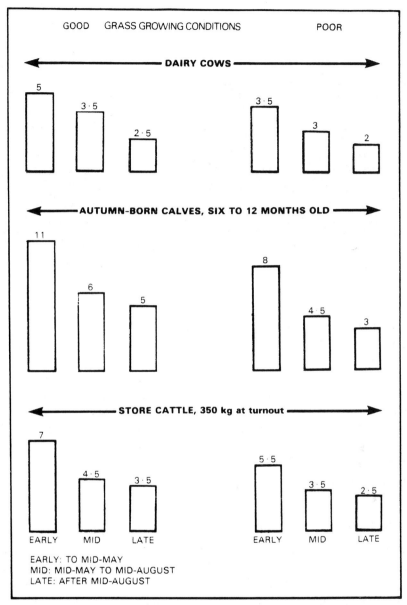

Fig. 4.4 Target stocking rates at pasture.

result, the feeding period is extended, feed costs are increased and margins are reduced. An example of this effect is shown in table 4.5. The target daily gain of 0.8 kg was not reached, but growth rate was still reasonable, at 0.65 kg/day. Weight at yarding was 30 kg lower, and because daily gain in the subsequent period was similar, at 0.9

Table 4.5 Consequences of failing to make target grazing gains in 18-month beef production.

| | Daily gains at grass | |
	0.8 kg/head	0.65 kg/head
Weight at yarding	325 kg	295 kg
Finishing period (at 0.9 kg/day weight gain in winter)	190 days	220 days
Extra silage needed		0.8 tonnes
Extra rolled barley needed		45 kg
Extra cost of feed (1982 prices)		
silage at £15/t		£12
rolled barley at £110/t		£ 5
Total		£17 per head

kg/day, the feeding period was thirty days longer than for those cattle which had met the target. The longer feeding period was reflected in an increased requirement for both silage and rolled barley, estimated to cost an extra £17 per head in winter-feed costs.

In contrast to dairy cattle, which appear to be relatively immune to gastro-intestinal parasitic infections, beef calves can suffer serious decreases in health and performance if care is not taken to control stomach worm (*Ostertagia ostertagii*) infestation of pasture. The critical period for calves is from mid-July onwards, when most of the eggs passed on to the pasture in dung in previous months have developed into infective larvae which are present on the herbage and are eaten by the grazing animals. The calves may then acquire large worm burdens which retard growth or cause clinical disease from August onwards.

It is financially worth supplementing dairy-bred calves and suckled calves in the latter part of the grazing season. Recommended rates of feeding and recorded responses are given in table 4.6. In the case of dairy-bred calves in the first season at pasture, feeding is worth while from mid-August onwards. Creep feed for suckled calves gives a useful response, but in this case it is best to supplement for a shorter period of time immediately prior to weaning, particularly with spring-born calves where milk is still likely to be a significant contributor to calf growth. Creep feeding has the advantage of conditioning the calves to future winter diets, and as a consequence the risk of a check in growth at weaning is likely to be reduced.

Table 4.6 Levels of supplement for beef calves in the second half of the grazing season.

	Supplement (kg/day)	Response (kg liveweight)
Dairy-bred calves (rolled barley)		
autumn-born (9–12 months old)	0.7	+ 9
spring-born (4–7 months old)	0.7	+12
Suckled calves (creep feed)		
autumn-born (9–11 months old)	1.0	+19
spring-born (8–9 months old)	1.0	+10

Lamb growth at grass

Traditionally the relative stocking rate of sheep has been much lower than that for beef or dairy cattle. Sometimes this reflects poorer grass growing conditions, but attempts to increase stocking rate even on good land have often been disappointing because increased sheep numbers have resulted in reduced lamb performance.

The main reason for the lack of improved output is the increased incidence of parasitic worm infestation, associated with keeping more sheep on the same land year after year.

The Meat and Livestock Commission[4] have calculated the financial reward from selling finished as opposed to store lambs (see table 4.7). Those flocks which produced mainly finished lambs received more money per lamb, and because their variable costs were similar to those flocks which produced store lambs for winter feeding, their gross margin per head averaged 14 per cent higher. Possibly the flocks which sold finished lambs were on better grassland, because they also had higher stocking rates than those which produced stores, so that their gross margin per hectare was 22 per cent higher.

Lamb growth can be severely depressed by worm infestations, principally the stomach worm *Ostertagia circumcincta*, and *Nematodirus* species. Eggs passed by ewes give a large increase in infective larvae of the stomach worm at the end of June, which cause poor growth or disease in lambs shortly after they have eaten the infected grass. Lambs are at risk from *Nematodirus* infection until they are four months old when they become progressively more resistant to infection. They are liable to infection only if the pasture was grazed by lambs the previous year. Ewes play a negligible role in the trans-

Table 4.7 Producing finished lambs rather than stores: effect on returns and gross margins.

| | Flocks producing mainly finished lambs compared to those producing mainly stores (1981 prices) | |
	(£)	(%)
Extra return per lamb	3.50	+ 8
Extra gross margin:		
per head	3.70	+14
per hectare	76	+22

mission of *Nematodirus* worms because they themselves are resistant to infection.

Supplementation of lambs post-weaning is likely to be money well spent if it accelerates growth and increases the proportion sold fat off grass. An appropriate rate of supplementation is 125 kg/head/day.

Clean grazing

The successful control of worms depends on having clean pasture available for grazing, especially in mid-July when the population of infective stomach worm larvae rises rapidly. Aftermath herbage, preferably not grazed at all during the spring before being closed up for hay or silage, is best. If the land was grazed before being closed up, it will remain safe if it was grazed by a different species of animal, or by animals which had been dosed with anthelmintic before being turned on to the pasture.

Clean grass can be provided by a three-year rotation of cattle, sheep and hay or silage (see fig. 4.5) — remember it by the mnemonic CASH. On farms where it is not possible to cut every field for hay or silage, cattle and sheep are alternated on an annual basis (see fig. 4.6). It helps in this version of clean grazing, if the total liveweight of the sheep and cattle populations are similar. In a third version (see fig. 4.7), which is practised at *Farmers Weekly* Curworthy Farm, the cows graze a separate area, and the sheep alternate with the silage area. This ensures adequate clean grazing both in spring and for the lambs after weaning.

The trick is to start the season with both clean animals, by dosing them at turnout, and clean pasture which was not grazed the previous year by the same animal species.

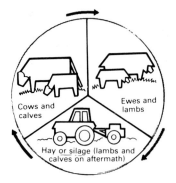

Fig. 4.5 The clean grazing system for sheep. The rotation is cattle, sheep, hay (or silage).

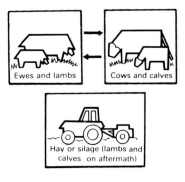

Fig. 4.6 Clean grazing of upland pastures by alternate sheep and cattle.

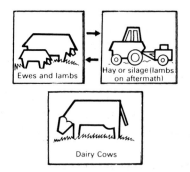

Fig. 4.7 The clean grazing system at Curworthy Farm; sheep alternate with forage conservation; weaned lambs graze aftermath regrowths.

Plate 4.3 Successful control of worms means starting the grazing season with clean animals and clean pasture which was not grazed the previous year by the same animal species.

If no clean pasture is available, calves should be dosed at three-weekly intervals until the end of May to prevent infection from larvae which have hatched from over-wintered eggs. This dosing regime will effectively clean up the pasture and prevent the mid-summer rise in infective larvae from occurring.

In the case of sheep grazing 'dirty' pasture in the first half of the season, the lambs will be infected with larvae by the time they are weaned. 'Worm and turn' is essential here: the lambs are turned on to clean pasture, but because they are wormed at the same time, they do not proceed to infect the clean land, which may be used for ewes and lambs next spring.

Results achieved from clean grazing at Edinburgh[5] are impressive (see table 4.8). Clearly the stocking rate and level of nitrogen must be related to the potential of the land, but the results demonstrate the relatively high output which can be achieved from clean grazing.

The two-pasture system for hill land

Work at the Hill Farming Research Organisation and at Redesdale EHF has shown the value of improving a proportion of land in conjunction with the open hill. Effort is first concentrated on the land which is most amenable to improvement, usually between 10 and 20 per cent of the total area.

Table 4.8 High levels of output can be achieved from clean grazing systems for sheep.

	Lowland	Upland
Breed/cross	Halfbred	Greyface
Stocking rate		
(ewes/ha)	17.5	12.7
(+lambs/ha)	30	19
Nitrogen (kg/ha)	210	150
Average growth rate of lambs (g/day),		
birth to weaning (120 days)	270	301
	(5-year average)	(3-year average)

The high growth rate of lambs was achieved despite the fact that the lambs were not dosed for worms during this period.

Plate 4.4 Improve 10 to 20% of the hill land which is most amenable. Graze it at lambing time and subsequently with ewes with twins. Flush ewes on it at mating time.

The strategy is to use the improved pasture, which has been re-seeded and fertilised, so that the sheep flock can derive most benefit from it. After lambing, ewes with twin lambs graze the improved pasture whilst those with singles are returned to the open hill. After weaning, all ewes are transferred to the hill. They then return to the improved pasture for flushing prior to mating (see fig. 4.8).

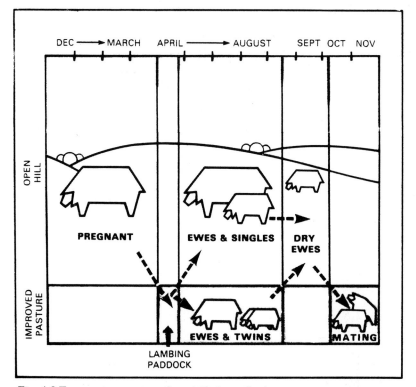

Fig. 4.8 Two-pasture system for a hill sheep flock.

The effect on output of introducing the two-pasture system can be spectacular. In the first six years after the introduction of 15 per cent of improved pasture, achieved by fencing, applying lime and slag, and grazing with cattle before oversowing with ryegrass and clover, lamb output at Redesdale EHF increased from 113 lambs from 155 ewes to 338 lambs from 372 ewes. This represents an increase in lambing percentage from 73 per cent to 91 per cent, and in addition the weight of lambs at weaning increased from 27 kg to 33 kg. At Pwllpeiran EHF in Wales, land improvement has now reached the point where larger Speckled-Face sheep have been introduced to replace the Welsh Mountain, with a corresponding increase in output.

References

1. Le Du, Y.L.P. and Hutchinson, M. (1982) in *Milk from Grass*, ICI/GRI, 44.
2. Illius, A.W. and Lowman, B.G. (1982) *Proceedings of the European Grassland Federation Meeting, Occasional Symposium No. 14*, British Grassland Society, 193—5.
3. Leaver, J.D. (1983) *Grass Farmer*, no. 14, 15—17.
4. MLC (1982) *Commercial Sheep Production Yearbook*.
5. Speedy, A. (1980) *Sheep Production*, Longman, 28—29.

5 Silage

The rapid growth of grass in spring means that by late May there is an excess of supply over demand which can be conserved for use during the following winter. Conservation thus solves two problems at the same time: it removes excess herbage which would otherwise be wasted, and thereby providing areas of young aftermath herbage for grazing in mid-season; and it also provides relatively low-cost winter-feed.

The main objective in making silage is to preserve the crop by fermentation with minimal loss of nutrients. The product should be sweet-smelling and capable of being eaten in sufficient quantity by the animal so that the required levels of output are achieved during the winter period. Predictability is the key to efficient silage making; without it the best laid plans can be completely thwarted. Some ways of achieving high-quality silage repeatedly are described in this chapter.

Quality or quantity?

The amount of land set aside for cutting affects both the quality and the quantity of silage; it also infuences the area available for grazing in early and mid-season.

On small farms it is traditional to let grass bulk up to yield as much as possible before being cut. It is argued that only then is there likely to be enough silage to meet requirements. But the longer the land is closed up the more area is required for grazing both in spring and later in the season. Grass regrowth is delayed to a greater extent following a late cut of a mature crop than when a younger crop is cut. This can lead to greater wastage of grass in the grazing areas.

Digestibility of grass at the time of cutting is the major factor determining silage quality. The D-value of ryegrass falls by about 2.5 units a week. Therefore the decline in digestibility during the silage period is of considerable significance.

Targets for quality need to be set, as well as for quantity. A

Plate 5.1 Conserving grass solves two problems at once: it removes excess herbage which would otherwise be wasted, and also provides a relatively low-cost winter feed.

realistic target is to aim for a silage with a metabolisable energy content of 10.7 (67 D-value) from a three-cut system. To achieve the required quantities of silage at this quality, the level of fertiliser nitrogen should be close to the optimum for the site class, and about 65 per cent of the total area of grass should be set aside for the first cut (see table 5.1).

The target total quantity of silage to meet the requirements of a cow yielding 6000 litres of milk, according to the ICI/GRI team[1] is just over 8 tonnes of fresh material per cow. The appropriate overall

Table 5.1   Target areas to be set aside for cutting silage in a three-cut system.

	First cut, mid-May	Second cut, end June	Third cut, mid-August
Proportion of total area to be set aside for cutting (%)	65	45	45

Under average grass growing conditions, about two-thirds of the total grass area should be cut in mid-May for silage.

Table 5.2   Milk from grass: targets for a three-cut silage regime.

ME of silage (MJ/kg DM)	10.7
Feed required for 6000 litres of milk,	
silage (t/cow)	8.1
concentrates (t/cow	0.9
Feed costs (silage and concentrates)	
(£/cow)	247
Stocking rate (average grass growing conditions)	
(cows/hectare)	2.0
UME from grass (GJ/ha)	95
Gross margin (£/ha)	1201

Despite the high requirement for silage and relatively low stocking rate, feed costs per cow are contained to give a relatively high target gross margin per hectare of £1200.

stocking rate for land in site class 3 (average grass growing conditions) is two cows per hectare (see table 5.2).

When to start cutting

Bearing in mind that digestibility declines during the silage-making period, it is advisable to start cutting a few days before the crop reaches the target D-value. Target starting dates for early, intermediate and late varieties of ryegrass are suggested in table 5.3.

It is a good idea to have fields sown to different varieties, so that the early varieties can be cut first, followed by the later ones. In this way, silage quality can be maintained over the whole silage-making period.

Table 5.3 First-cut silage: when to start cutting.

	Target starting dates for a 3-cut system to give silage of 10.7 MJ ME
Early ryegrasses (e.g. Cropper, S24, Frances, Reveille, RVP, Sabalan)	12 May
Intermediate ryegrasses (e.g. Talbot, Combi)	19 May
Late ryegrasses (e.g. S23, Meltra, Melle, Endura)	25 May

Later flowering varieties can be cut 10 to 12 days later to give silage of similar yield and ME value.

Delayed cutting of the first-cut silage crop has the following consequences:

(a) increased risk of grass shortages in mid- and late season;
(b) silage of low ME value;
(c) increased need to supplement the silage to achieve high performance in winter.

There is little point in growing the crop on to the stage where more has to be harvested, carted, stored, fed out, eaten, then carted to the slurry lagoon. Indigestible silage is a waste of money, especially if you have too much of it.

Which harvesting system?

A silage-making system, however small, needs to be planned carefully if it is to conserve high quality forage. If the bulk of the crop cannot be conserved by the time 64 D is reached (i.e. in ten days) then it might pay to look critically at the existing system and consider whether output per day and per season is as high as it might be.

Work rates for forage harvesters are not very well established; a do-it-yourself approach is shown below:

$$\text{Hectares/hour} = \frac{\text{Speed (km/h)} \times \text{effective working width (m)}}{\text{13}}$$

Your own figure compared to that given by the manufacturer, which represents the potential work rate, will indicate any bottlenecks in the system, such as inadequate trailer capacity or slow filling of the silo. If, however, output is reasonably high and the bulk of the grass is still not being harvested at the optimum D-value then a higher capacity system is needed.

The costs of making silage by seven contrasting systems are detailed in table 5.4. The probable work rate, or output per hour is lowest for the two-man flail harvester system, which is also the cheapest. The high-density big baler has the highest work rate, but since it is a relatively new piece of equipment there is less information on its performance from field trials than for the other systems. There is an urgent need, however, for up-to-date independent information on the performance in the field of all types of forage harvesters.

The calculations given in table 5.4, based on the existing limited information on the work rate of forage harvesters, show that the two-man flail harvester system can produce enough silage for a 65-cow herd, whilst the forage wagon, big baler and double-chop systems can cope with silage for up to 140 cows. The metered-chop harvester can produce enough silage for up to 260 cows, whilst the self-propelled harvester and the high-density big baler are suitable for making silage for large herds.

Clearly there are large effects of scale of operation on the cost per tonne of DM ensiled, particularly with the two high-performance systems. An investment of £50,000 in equipment, together with the need to employ six men in the case of the self-propelled harvester make these two systems particularly well suited to the contractor. The big baler compares favourably with both the forage wagon and the double-chop harvester. This reflects the fact that it works with a higher dry-matter crop, since the fixed and variable costs of operating the equipment are very similar. A bonus to the big bale however, is that storage costs per tonne are almost halved (£3.80/tonne fresh weight) compared to the other systems which involve using a conventional silo (£6.80/tonne fresh weight).

Short chopping, as occurs with metered-chop machines, has the advantage of increasing trailer and silo capacity, and of releasing sugars more extensively to give a more rapid fermentation of the crop when it is in the silo. But it requires more tractor power than double-chopping or harvesting by forage wagon or by big baler.

For the 100-cow herd, then, the big bale system offers scope for cost-saving, particularly if silo capacity is limiting. It is vital, though,

Table 5.4 How much does it cost to make silage?

Harvester	Flail	Forage Wagon	Big Baler	Double-chop	High-Density Big Baler	Metered-Chop	Self-Propelled
Men	1	2			3	5	6
Probable output (ha/h)	0.21	0.5	0.5	0.5	1.8	0.8	1.25
Maximum capacity (ha)	30	72	72	72	260	120	180
Cows	65	140	140	140	570	260	400
Capital cost of equipment (£)	5,000	10,000	12,000	12,000	49,000	16,000	56,000
Total cost (£/tonne DM)†							
for 30 ha of crop	73	54	37	60	*	72	*
maximum capacity of system	73	41	26	45	25	46	55

* Contractors' operation at lower areas.
† 20% DM for flail, 25% for double and metered-chop and 35% for forage wagon and big balers.

Plate 5.2 Harvesting wilted grass by metered-chop harvester. Short-chopping increases trailer and silo capacity and promotes a rapid fermentation in the silo.

that losses are controlled by complete sealing throughout the period of storage.

To wilt or not to wilt?

There is little benefit from wilting, despite higher intake, when comparisons are made between well-preserved wilted and direct-cut silages made from the same initial crop of grass. This reflects similar total losses from cutting to feeding (see table 5.5.).

The main advantages of wilting are increased speed of harvest, and reduced production of effluent from the silo. Wilting to 25 per cent dry matter, when it is impossible to squeeze juice from a handful of chopped grass, means that fewer trailer loads are required to harvest a hectare of crop, since 30 per cent less water is being harvested. The

Table 5.5 To wilt or not to wilt? Well-preserved direct-cut silage compared to well-preserved wilted silage.

	Direct-cut (18% dry matter)	Wilted (25% dry matter)
Field		
Trailer loads to harvest one hectare (5 tonnes dry matter yield)	10	7
Water harvested (tonnes/ha)	28	20
Silo		
Effluent production (litres/tonne)	100	20
Total loss of dry matter (%)	20	20
Animal		
Intake	lower	higher
Digestibility	higher	lower
Liveweight gain	similar	
or		
Milk yield	similar	

The aim should be to wilt for no longer than 24 hours with the objective of ensiling at 25% dry matter.

additional job of mowing in advance of the harvester can be done speedily and normally would not require extra labour or an additional tractor. A wilted crop ensiled in a bunker silo should produce relatively little effluent and at a steadier rate than with direct-cut crops.

Loss of nutrients via respiration in the field may result in a slightly lower digestibility value for wilted compared to direct-cut silage, but to balance this the level of intake of the wilted product is likely to be higher when both are offered to the animal *ad lib*. Thus animal performance is unlikely to be very different when both silages are well-preserved. If, however, the direct-cut silage is badly preserved, then the wilted crop is likely to support higher levels of output.

Direct-cut crops may be ensiled on top of straw in an attempt to reduce effluent loss. There may also be benefit to the straw from the soaking it receives, apart from the nutrients contained in the effluent. It would probably be necessary to have a two-course layer of straw bales in a conventional bunker silo. The straw may be expected to absorb about twice its own weight of effluent.

Dried sugar-beet pulp may also be used in the silo to soak up some of the excess moisture from wet crops. The advantage of this material is that it also supplies fermentable sugar, though the extent to which this extra sugar may benefit the fermentation will depend on the amount of pulp added to the grass, and the sugar content of the product.

When to use an additive

Additives are used to ensure efficient preservation. Under good weather conditions grass crops should ensile satisfactorily without the use of acids, salts or additional sugar.

Additives with formalin, giving a degree of protein protection, or formic acid, may still prove advantageous even when wilting is practised, although an economical return has not yet been clearly demonstrated.

A valuable contribution has been made to this topic by the team at Liscombe EHF.[2] They devised a star-rating system for assessing the risk of poor preservation which took into account a number of factors in addition to the weather. The Liscombe scheme is shown in table 5.6 in a modified form.

The principle is that sugar content is the critical component in determining the pattern of fermentation; if there is insufficient sugar a secondary fermentation is likely to occur. This means that although lactic acid may be produced in the first 10–14 days after ensiling, there is insufficient acidity to prevent the growth of the clostridial bacteria. These organisms are relatively intolerant of pH levels below 4.5. But if they do start to grow, they ferment lactic acid to evil-smelling butyric acid, and also degrade proteins and amino acids to ammonia. Silages with more than 15 per cent of total nitrogen in the form of ammonia nitrogen have undergone a secondary fermentation and are classified as being poorly-preserved.

The higher scores in the Liscombe table indicate higher concentrations of sugar in the crop. The target level of sugars, above which most crops would be expected to ferment well, is 3 per cent of the crop fresh weight. Ryegrasses have higher contents of sugar than other grass species or the legumes; water content decreases and sugars accumulate as digestibility decreases with advancing crop maturity; fertiliser-N tends to increase leafiness and decrease the content of sugar in the crop.

Forage harvesters can influence the pattern of fermentation in two ways. Those which chop the crop short also release the plant sugars to a greater extent, thereby promoting a more rapid fermentation.

Table 5.6 When to use additive: the Liscombe scorecard.

	5	4	3	2	1	Your Score
			SCORE			
Species	Italian ryegrass maize	Perennial ryegrass	Other grasses or grass and clover		Legume	
D-value		less than 60	60–65	more than 65		
Fertiliser-N kg/ha per cut			less than 50	50–100	more than 100	
Forage harvester	Metered-chop	Double-chop	Flail	Forage wagon		
Weather	Sunny		Cloudy		Showery	
Season			Spring and Summer		Autumn	

Total score ====

Score	Risk of poor preservation	Additive need
Above 20	low	Not normally required
15 to 20	medium	Recommended rate
Below 15	high	High rate of addition

Secondly, harvesters tend to act as inoculation agents by virtue of the fact that grass juice sticks to the inside of the machine. This juice harbours lactic acid bacteria.

Sunny weather not only means higher levels of sugar in the crop resulting from photosynthesis, it also means that less water is present at the time of cutting.

Autumn grass is invariably more difficult to ensile well than spring grass, principally because sugar levels are likely to be relatively low and the crop is likely to be leafy and relatively wet.

Examples are shown in table 5.7 of how different crops and
harvesting conditions might score. Italian ryegrass (Crop 1), harvested
at 65 D-value after receiving 80 kg N for the cut by metered-chop
harvester in cloudy spring weather, scores well and should not
require an additive to assist preservation, which should be good. On
the other hand, the grass/clover crop (Crop 2) harvested at 62 D by
forage wagon in showery autumn weather scores badly and would
probably require double the recommended rate of effective additive.

Table 5.7   Examples of how your crop may score.

Crop 1	Score	Crop 2	Score
Italian ryegrass	5	Grass + clover	3
65 D-value	3	62 D-value	3
80 kg N/ha per cut	2	Zero N	3
Metered-chop		Forage wagon	2
forage harvester	5		
Cloudy weather	3	Showery	1
Spring	3	Autumn	1
Total	21	Total	13
Additive not required		Use additive at high rate	

Which additive?

Making silage is like making beer or wine at home; if the rules are
followed carefully, a reasonable product can be made every time.
Additives are designed to increase the predictability of achieving a
well-preserved silage.

The ADAS assessment[3] of the various different types of silage
additive is reproduced in table 5.8. When selecting an additive
efficient preservation and improved animal performance are the main
factors to be considered, but aspects relating to application are also
important. The star ratings given by ADAS for effectiveness relate to
the active ingredients at the highest rate of application recommended
by the manufacturer.

Animal performance should be improved if the additive has been
effective in improving preservation quality, because losses of

Table 5.8 The ADAS assessment of groups of silage additives.

| | | Animal Performance | | | | | |
	Preservation	Dry matter Intake	Utilisation	Protein Protection	Application	Minimum Corrosion	Remarks
Acid/formalin	*****	***	****	***	**	***	Inorganic acids less effective than organic. Most useful at higher DM levels, i.e. above about 20%.
Organic acid	*****	***	***	*	*	*	Most effective with wet materials.
Inorganic acids	**	0	0	0	**	**	Acid effect only. Recommended application rates often too low.
Acid mixtures	*	0	*	*	***	****	Level of active ingredients and application rate often too low.
Salts	**	0	*	0	****	*****	Not always effective at low levels of application. Limited information suggests tetraformate similar to formic acid.
Sugars	****	0	0	0	****	*****	High levels required to be effective.
Inoculants	I	I	I	0	*****	*****	Need further investigation under UK conditions.

Coding

*****	Excellent
****	Very good
***	Good
**	Fair
*	Poor
0	No effect expected or observed
I	Insufficient information available to date

digestible nutrients will have been lowered and in *ad lib* feeding situations the animal will most likely eat more of the ensiled product. Protection of protein from extensive degradation in the rumen may be conferred by ingredients such as formaldehyde which form chemical bonds with the proteins in the crop.

There is insufficient independent information for an assessment of the likely improvements in performance by animals given the crops treated with the different additives. Declaration of ingredients would clearly help in predicting the likely effects of additives on the quality of preservation and on feed value. Unfortunately the composition is stated on only about half the products available on the market. The 'Wilkinson Guide to Silage Additives', based on the ADAS booklet, lists a few example products which offer value for money (table 5.9). The best buy is formic acid. It is well-proven and the cheapest product at the recommended rate of active ingredient.

Formic acid and acid/formalin mixtures have consistently given improvements in milk yield in a limited number of comparative trials with dairy cows (fig. 5.1). The response averaged 1 kg extra milk per day over a wide range of yield levels. With beef cattle, formic acid has been comprehensively tested and has given improved performance over a very wide range of liveweight gains (fig. 5.2). In all these trials, the same crop was ensiled either with or without additive and then given to similar groups of animals *ad lib*. When the diet included supplements they were given at the same level of feeding. Each dot in the two figures represents one comparison between untreated and additive-treated silage.

Acid mixtures, according to the ADAS booklet, are something of an unknown quantity. They are dismissed in table 5.8 as being 'poor' on preservation, with the added remark: 'level of active ingredients and application rate often too low'. Since virtually every product is an unknown quantity with regard to composition, it is not possible to calculate the application rates of the active ingredients, and it is therefore impossible to know whether they would offer value for money if added at rates equivalent to those recommended in table 5.9. Acids and acid mixtures score poorly on attributes associated with application. Some can remove paint from machinery and thereby make them more susceptible to rust damage.

Salts of acids are an attractive alternative to straight acids, particularly if they have similar properties. Of the products listed in the ADAS booklet, only ammonium tetraformate, which appears to be similar to formic acid, is cost-effective at the recommended rate of addition.

Table 5.9 The Wilkinson guide to silage additives.

Group	Recommended active ingredient	Recommended rate of addition (active ingredient as % of crop fresh weight	Example Product	Recommended rate of addition (litres/tonne fresh crop)	Cost at recommended rate of addition (£/tonne)
Acid	Formic acid	0.25	Add-F	2.5	1.22
Acid mixtures	Formic acid + Propionic acid	0.20 + 0.10	Insufficient information available		
Acid + formalin	Formic acid + Sulphuric acid + Formaldehyde	0.13 + 0.18 + 0.04	Farmline	4.0	1.64
Salts	Ammonium tetraformate	0.30	Foraform	3.5	1.78
Sugars	Molasses	1.0	Molasses	15	1.35
Inoculants	Lactobacillus plantarum + Streptococcus thermophilus	Insufficient information available	Insufficient information available		
Enzymes	Cellulases + Hemicellulases	0.20	Insufficient information available		

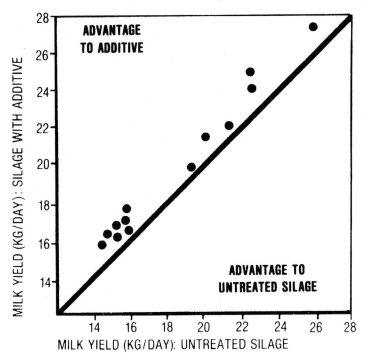

Fig. 5.1 Milk yield by cows given silage made from the same crop either without additive or with addition of formic acid and/or formalin.

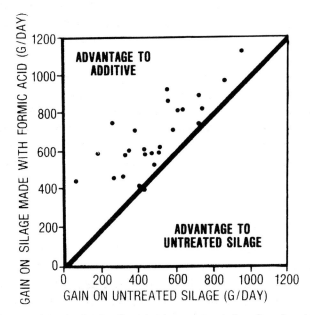

Fig. 5.2 Liveweight gains by beef cattle given untreated or formic acid-treated silage (g/day).

Molasses is cheap, even when added at the recommended rate of 1 per cent of the crop fresh weight, but its viscosity is a limiting factor to its use, coupled with the fact that 1 per cent represents 70 litres of this dense liquid per tonne of crop. This is a very high rate of application and may limit the speed of harvest.

There is now a new generation of additives, the bacterial inoculants, which have yet to be fully assessed under UK conditions. Ideally, they should boost the population of lactic and acid-producing organisms substantially, and should reduce losses to the extent that the feed value of the treated silage is greater than that of the untreated crop. Inoculants, which promote rapid fermentation, and which suppress oxidation in the early stages of ensiling may be useful, particularly with high dry-matter silages.

Whichever product is chosen, its utility ultimately depends on realising a benefit which is greater than the cost of the additive. Two possible benefits may accrue: a reduction of losses during ensilage, and an increase in animal performance. Studies have shown that losses can be reduced following the addition of formic acid, from about 25 per cent to 20 per cent. If silage is valued at £80 per tonne of DM this saving is worth £4 per tonne of DM — less than the cost of the additive (say, £1.50 per tonne of fresh crop, or £6 per tonne of DM). On the other hand, an increase in D-value of 2 units coupled with a 10 per cent increase in feed intake will supply enough metabolisable energy for 2.4 litres of milk. The extra silage would cost 6.8p and the milk would be worth 36p — a cost-benefit ratio of 1 to 5 (see table 5.10).

Although it is undoubtedly useful to have more silage to feed as a result of reducing losses by using an effective additive, the real bonus is in terms of improved animal performance. This effect has been illustrated in trials with growing beef cattle, where marked differences in performance have been seen between effective and ineffective additives when both have been challenged under difficult ensiling conditions.

How to minimise losses

'The experts' generally talk in terms of dry matter loss. This is only part of the story, since farmers deal with the whole crop, water and all. Further, dry matter loss is not the same as energy loss as anyone who makes beer or wine knows. Here, yeasts ferment sugar to alcohol and in the process half the dry matter is lost as carbon dioxide and water; but hardly any energy is lost from the system. The

Table 5.10   The cost-benefit ratio of silage additives.

Assumptions

Value of milk   = 15p a litre
Value of silage   = 8p/kg dry matter (£80/tonne)
Cost of additive   = £1.5/tonne (£6/tonne of dry matter)

Reduction of loss
If additive gives a reduction of DM loss of 5 per cent, the extra silage DM is worth £4/tonne.

Increased milk production
If additive gives an increase in D-value of 2 per cent and an increase in intake of 10 per cent, this will supply enough metabolisable energy for 2.4 litres of milk a day.

Cost of extra silage   =   6.8p
Value of milk   =   36p

---

product — beer or wine — is quite fattening as a result. Silage tends to be a slightly more concentrated form of feed energy than the fresh grass from which it came; it has lost dry matter but not quite as much energy during storage.

These aspects were discussed on a large dairy farm, where they had experienced difficulty in wilting the crop. Fortunately, they had produced a well-preserved silage by using an effective additive. The farmer was shocked to learn that although dry matter loss was likely to have been quite low, the fresh weight loss was estimated to be as high as 40 per cent. He immediately set about purchasing additional bulk feed to ensure he had an adequate total supply of food for the winter.

He then worked out what might have happened, from the estimated DM content of the crop at harvest and the silage analysis. It was suggested, after inspecting the silage and the silo, that a total DM loss of 20 per cent was reasonable for well-preserved direct-cut silage. The calculations are shown in table 5.11. The apparent increase in DM content of 6 per cent between harvesting the crop and analysing the silage most likely reflects the fact that water loss in effluent exceeded DM loss from this fairly large bunker silo. Thus, fresh weight losses of 40 per cent can be quite costly, and should be anticipated along with the more normal 20 per cent dry matter loss.

Some sources of loss are unavoidable; others can be virtually

Table 5.11   Losses of fresh weight can be substantial even with well-preserved direct-cut crops.

Fresh crop harvested	1500 tonnes
Dry matter of fresh grass (estimated)	18%
Dry matter harvested	270 tonnes
Assume 20% loss of dry matter during storage	
Silage dry matter available for feeding	216 tonnes
Dry matter of silage (from analysis)	24%
Fresh weight available for feeding	900 tonnes

$$\textit{Loss of fresh weight during storage} = \frac{1500 - 900}{1500} \times 100 = 40\%$$

eliminated. The range of energy loss from each source is shown in table 5.12, taken from studies in the Federal Republic of Germany.[4]

Loss due to respiration during field-wilting will be about 2 per cent per day in the field. Flail mowers and mower-conditioners can accelerate removal of water from the crop and reduce the time required to reach the target DM content; but if rain occurs then loss due to leaching can be greater than with unlacerated or unconditioned crops. The alternative approach is to harvest the crop directly after mowing and deal with the effluent from the silo. The main determinant of effluent is, of course, the amount of water in the crop, but increased depth of silage in the silo, short-chopping and the use of acid additives can all lead to increased effluent production.

Plant respiration in the silo occurs from the time the crop enters the silo until either the pH has fallen sufficiently to inactivate the plant enzymes, or the supply of oxygen is exhausted. Loss from this source can be minimised by short-chopping, consolidating the crop – which can prove difficult if the crop is over-wilted – and sealing the silo as soon as possible. Formic acid reduces respiration by virtue of the fact that it decreases the pH of the crop at the outset to below 5.0.

Fermentation is a relatively efficient process in terms of conserving

Table 5.12   Energy losses during ensilage.

Source	Loss (%)	Influenced by
Unavoidable		
Respiration during wilting	2 to> 5	Weather, duration of wilting period, type of crop, type of mower
or		
Effluent from the silo	5 to > 7	Dry matter content, depth of silo, chop length, additive
Respiration in the silo	1 to 2	Dry matter content, chop length, (additive), sealing
Fermentation	2 to 4	Dry matter content
Avoidable		
Secondary fermentation	0 to > 5	Type of crop, dry-matter content, additive
Surface deterioration during storage	0 to > 10	Speed of filling, density, sealing, type of silo
Surface deterioration during feed-out	0 to > 15	Density, type of silo, unloading technique, season
Total	8 to > 40	

energy, though losses from secondary, clostridial, fermentations may be as high as 15 per cent.

Losses due to surface deterioration can be virtually eliminated by complete sealing, especially at the shoulders, and by having a dense mass of silage which resists the penetration of air.

The values in table 5.12 indicate that under exceptionally good management, losses may be contained to below 10 per cent. On the other hand, badly made silage may lose over 40 per cent of its energy to the atmosphere and appear more like a compost heap than a silo. Over-heating, a consequence of inadequate sealing, gives a brown product which may smell like caramel or tobacco. When this occurs there has been severe damage to the protein fraction, and it is unlikely to be readily available to the animal.

In table 5.13, typical values are shown for losses from direct-cut and wilted silages made under good management.

Plate 5.3 Losses can be reduced by consolidating and by complete sealing and by protecting the seal from damage by wind, birds and vermin. Where tyres are used they should be placed as close together as possible.

Table 5.13   Typical losses of dry-matter in silage making.

Loss (%)	Direct-cut (with formic acid)	Wilted (24 h)
In field		
Respiration	—	2
Mechanical loss	1	4
During storage		
Respiration	—	1
Fermentation	5	5
Effluent	6	—
Surface waste	4	6
During removal from store	3	3
Total	19	21

What to look for in silage analysis

The really useful measures of silage value are dry matter, meta-bolisable energy (ME) and NH_3-N. Dry matter indicates how much potentially useful feed is in the clamp, and how much water is there. The content of ME shows how useful the dry matter is as a source of energy to the animal. Ammonia nitrogen (NH_3-N) is an index of fermentation quality. The pH value indicates the extent of fermentation — how acid the silage is, whilst crude protein is of relatively little value since about half the so-called crude protein in silage is in the form of degradation products such as amides, amines, amino acids and ammonia.

The important measurements are summarised in table 5.14. In addition, the silage should smell sweet and taste sweet.

The ten commandments of silage making

1. Thou shalt not wilt for longer than 24 hours.
2. Thou shalt chop the crop reasonably short.
3. Thou shalt apply an effective additive as necessary.
4. Thou shalt avoid soil contamination.
5. Thou shalt fill the silo as rapidly as possible.
6. Thou shalt cover the silo at the end of each day's filling.

Table 5.14 Silage evaluation: important measurements.

Dry matter (DM)	Percentage or g/kg fresh weight. Usually indicates degree of wilting, but in direct-cut crops or crops made in wet weather it can be higher than expected because of effluent loss.
pH	Indicates extent of fermentation. The lower the value, the more acid the silage. Can indicate quality of perservation, but not as good as ammonia-N. Wet crops are unlikely to be well-preserved if pH is above 4.5. In dry crops, fermentation is restricted by lack of water, and pH can remain above 4.5 despite the crop being well-preserved.
Ammonia-nitrogen (NH_3-N)	Percentage or g/kg total nitrogen. Indicates quality of preservation. Below 5% very good; 5–10%, good; 10–15%, moderate; above 15%, poor. Should be measured on silage which has been in store *at least 3 months*.
Metabolisable energy (ME)	Megajoules (MJ) per kg DM. Indicates energy value. Now replacing D-value. Predicted from MAD fibre or D-value.
Crude protein (CP)	Percentage or g/kg DM. Indicates total nitrogen. Not of great use since only about half the N in silage is in the form of protein.

7. Thou shalt seal completely, checking for holes and taping them over.
8. Thou shalt protect the sheeting from the elements, the birds of the air and vermin that can chew plastic up.
9. Thou shalt keep a smooth silage face during feed-out; remove 10 cm per day in winter, 30 cm per day in summer.
10. Lest the Water Authority take action against thee, thou shalt not pollute rivers and streams with silage effluent. Collect effluent, dilute it and spread it evenly on thy land.

References

1. *Milk from Grass* (1982), ICI/GRI.
2. Liscombe EHF (1981) *Grass Bulletin*, No. 2.
3. *UK Silage Additives* (1983), MAFF/ADAS.
4. Zimmer, E. (1980) *Occasional Symposium No. 11*, British Grassland Society, 186–97.

6 Hay

High-quality grass crops are difficult to wilt and difficult to make into hay. But well-made hay is extremely good value for money as a feed. It has the additional advantage of being readily saleable off the farm.

Enthusiasts for hay place great stress on speed, so that the hay-making system is less dependent on the weather. Particularly critical is the time spent in baling and carting the hay. The objective is to bale and remove the crop from the field at a similar rate to that achieved by a metered-chop forage harvester in a silage system. The reason for stressing speed is not only because good weather can deteriorate quickly but also because the rate at which the crop loses water in the later stages of drying is very much slower than in the early stages.

Water loss in the field

On a good summer day a grass crop loses about twice its own weight of water by transpiration. Since it takes much longer than a day for a hay swath to dry, there is resistance to water loss in the plant and in the swath. This resistance can be reduced by mechanical conditioning of the crop at cutting, and by turning the swath often in the later stages of drying.

As the moisture content decreases, the rate at which the swath loses water decreases markedly, particularly between 33 per cent and 20 per cent moisture content, the difference between baling for barn drying and baling field-dried material. In the early period of drying, water loss is rapid and continues through the night. But towards the end of drying the swath can gain moisture from the atmosphere at night, which counterbalances the loss during the day (see fig. 6.1).

Dry matter loss during drying

Not surprisingly, losses of dry matter in the field are higher during

Plate 6.1 Towards the end of drying the rate of water loss is slow and the swath should be turned frequently at this stage.

the final phase of drying, especially in poor weather conditions (see fig. 6.2). It is important to realise the great effect of rain on nutrient loss. Studies in the Federal Republic of Germany[1] have shown that field-dried hay made in three days of good weather is likely to suffer a loss of DM of only 12 per cent. Under poor weather conditions the loss can be almost three times as high and digestibility will be several units lower as a result.

Thus the benefits of speeding up the rate of drying can be great, particularly if it means that the hay crop can be 'won' before a break occurs in the good weather.

Conditioning

The effect of conditioning on drying rate has been studied extensively at the NIAE, Silsoe.[2] Initial trial results are shown in fig. 6.3. The increase in dry matter content was greatest when the crop was both conditioned and turned daily.

The original machines had V-form steel spokes which abraded the surface of the crop. Despite the effectiveness of the original concept,

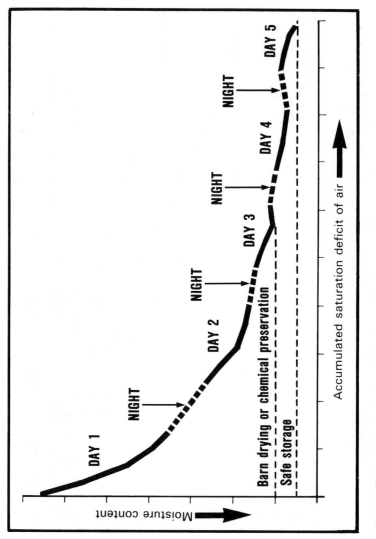

Fig. 6.1 Rate of drying in the field of a swath of perennial ryegrass.

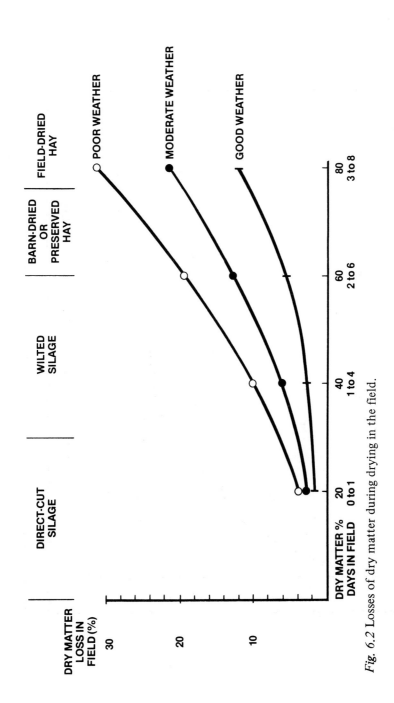

Fig. 6.2 Losses of dry matter during drying in the field.

design improvements have provided a constant challenge. It was desirable to reduce the cost and weight of the conditioning rotors and minimise the risk of damage to subsequently-used machinery, should the metal elements become lost. As a result, the latest V-form elements are made of plastic.

As the NIAE programme developed, the benefits of crop surface abrasion, rather than severe laceration, became increasingly apparent. The next step was to replace the V-form element by plastic tufts, which soon developed into the plastic brush. Then twin brush rotors were introduced (see table 6.1) which gave substantially greater effectiveness compared to a single brush. The most recent designs comprise serrated plastic ribs instead of rows of tufts; they are simpler, cheaper and likely to be effective at lower rotor speeds.

But all the twin-rotor machines have the disadvantage of a high power demand. Research is now concentrating on developing conditioners which are mounted on the drum of the mower. The objective is to make mower-conditioners more compact and simpler and better-suited to being mounted at the front of the tractor. Another conditioner, or a swather can then be mounted at the rear to give a

Plate 6.2 A serrated rib conditioner for hay-making developed by the National Institute of Agricultural Engineering.

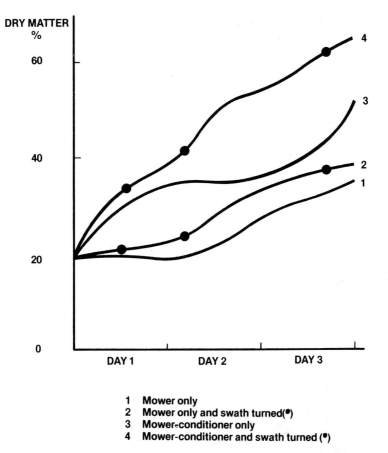

Fig. 6.3 Effect of conditioning on the drying rate of a hay crop.

really high-capacity outfit. The trick is to drive the cutting blades at 1200 to 1500 r.p.m. and the drum with serrated conditioning ribs at a quarter that speed, to save energy.

When to cut for hay

Apart from the susceptibility of hay crops to adverse weather during drying, much hay is made at a mature stage of growth and is of lower feed value than silage crops (see table 6.2). The objective should be to maximise hay quality by reducing the risk of damage by rain, and also by cutting as leafy a crop as possible. The Meteorological Office now offers a detailed weather forecasting service for farmers, which comprises estimating the probability of rain for consecutive three-day periods.

Table 6.1   Increases in drying rate following conditioning with NIAE crop conditioners.

	Increase in drying rate (%)
Single rotor: Steel V-form	50
Plastic V-form	53
Brush (with brush concave)	49
Serrated rib	89
Twin co-rotating rotors: brushes	135

Increase in drying rate compared to unconditioned crop, tests made under constant laboratory conditions.

Table 6.2   Average feed value of hay compared to silage: samples analysed by ADAS from farms in England and Wales.

	Hay	Silage
Metabolisable energy (MJ/kg dry-matter)	8.9	10.1
Crude protein (% of dry-matter)	9.6	14.4

In addition to obtaining weather information before starting to cut, it is a good idea to have a programme for cutting, turning and baling so that not all the crop is at risk at the same time. For example, a third of the total area could be cut and then turned. If the weather stays dry a second third can be cut on the third day, and so on.

Leafy hay can be made by cutting earlier in the season. But this means making hay when the probability of dry weather is lower, and when day length is shorter than in mid-June. Farmers with livestock often graze their hay fields early in the season before closing them up for hay. This has the effect of delaying stem development and ear emergence, which in turn delays the decline in feed value. To avoid excessive damage by poaching in early spring it is better to graze hay fields with sheep rather than cattle. Fields should be grazed until mid-April then closed up for cutting for hay in early June.

Grasses for hay

Another interesting recent development is the discovery at the Grass-land Research Institute, that tall fescue loses water at a much faster rate than other grasses. In very good weather, both unconditioned and conditioned swaths of S170 tall fescue dried to 33 per cent moisture content twice as fast as comparable swaths of Cropper perennial ryegrass (see table 6.3). The crops were grown in the same field and cut at the same time. Both species were at ear emergence.

Table 6.3   Tall fescue dries twice as fast as ryegrass.

	Relative drying rate to 33% moisture (unconditioned ryegrass = 100)	
	Unconditioned	Conditioned
Ryegrass	100	71
Tall fescue	52	33

In this trial the effect of crop species and of conditioning were additive; that is, the use of the conditioner has as great an effect on the fescue as it did on the ryegrass. Comparing unconditioned rye-grass with conditioned fescue, there was a three-fold difference in drying rate.

Ryegrass at ear emergence has its stems well-protected by leaf sheaths, and is a particularly difficult crop from which to make hay. Tall fescue, by contrast, not only has more pores (stomata) in its leaves, but its stems are more exposed and thus able to lose water at a faster rate.

Preservatives for moist hay

On the available evidence, the chemicals most closely satisfying the ideal specification are the ammonium salts of propionic or butyric acid. This conclusion is supported by the team of chemists and microbiologists at Rothamsted Experimental Station who have screened hundreds of chemicals for their effectiveness as hay pre-servatives. It is also reflected in the products currently on the market.

The most widely used hay preservatives contain either propionic acid or ammonium salts of propionic acid. Unfortunately many of the products are cocktails of unknown specification, so it is not always possible to calculate the amount of recommended active ingredient likely to be retained in the hay bale following application of the product (see table 6.4). A further feature is that in almost every case the recommended rate of addition is unlikely to be sufficient to give adequate preservation.

The objective of using a preservative is to prevent the development of pathogenic moulds and to reduce nutrient losses by controlling heating during the early period of storage. It is generally accepted that if the hay can be prevented from heating above 35°C, then it is likely to be adequately preserved, though sometimes there are anomalies; thermophilic moulds have been found in hays where excessive temperatures have not been recorded and vice versa.

Clearly, the wetter the hay the greater the risk of heating and moulding, and the more preservative is required to control deterioration. Guideline levels of *retained* propionic acid (or its equivalent in a salt) in the bale are shown in fig. 6.4 in relation to the moisture content of the hay, which can be estimated with some degree of accuracy by using a modified grain moisture probe in the bale. Uni-

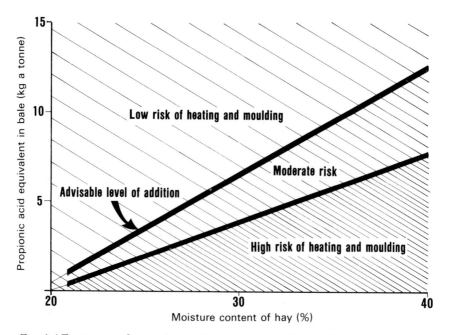

Fig. 6.4 Treatment of moist hay with preservative before baling; required quantity of propionic acid equivalent in the bale, in relation to moisture content at baling.

Table 6.4 The Wilkinson guide to hay preservatives.

Group	Recommended active ingredient	Recommended rate of active ingredient retained in bale; crop of 30% moisture (kg/tonne)	Example product	Recommended rate of addition* (kg/tonne)	Cost at recommended rate of addition (£/tonne)
Acid	Priopionic acid	6	Haycare	11	10
Salt	Ammonium bis-propanoate	7	Add-H	12	7

* Assuming 30% loss of product on application.

formity is vital to the success of the operation; the swath should have a uniform moisture content and the preservative should be distributed as evenly as possible within the bale.

The usual point of application of preservative is above the baler pick-up. Other options are to apply the preservative either on to the swath immediately prior to baling, or in the rowing-up operation immediately prior to baling. Trials by ADAS have shown that a dribble bar can be used successfully to apply preservative to the swath. The bar is mounted at the front of the tractor pulling the windrower, and the preservative is pumped at low pressure in large droplets on to the spread crop. Losses of chemical were estimated to average 30 per cent using this method. It is important to have the baler following immediately behind the applicator so that the target application rate can be checked by weighing the bales and measuring the quantity of preservative they have received. Baling straight after application also minimises loss of preservative to the atmosphere.

Feed value of hay made with preservative

Trials at Drayton Experimental Husbandry Farm with beef cattle have shown that provided the preservative is evenly distributed in the hay, and the rate of addition is adequate, then the feed value of the treated material is likely to be very similar to that of the same hay preserved by barn-drying (see table 6.5).

Bale handling

Bale handling is one of the major problems in haymaking and second in significance to the weather risk. A summary of a three-year survey of handling systems, carried out by ADAS[3] is shown in fig. 6.5. Improving the level of mechanisation by using mounted or trailed

Table 6.5   Feed value of hay made with a preservative or barn-dried.

	Barn-dried	Preservative*
Intake of dry matter (kg/day)	6.94	6.89
Liveweight gain (kg/day)	0.87	0.90

* Ammonium bis-propanoate added at 15 kg/tonne to hay of 24% moisture content at baling.

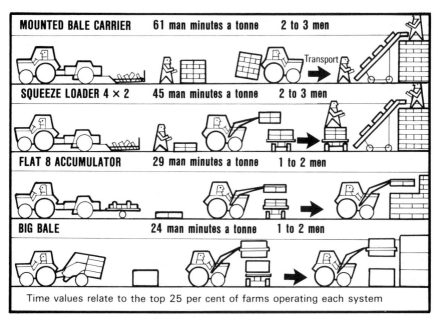

Fig. 6.5 Bale handling methods compared.

Plate 6.3 Bale handling: a flat eight accumulator and squeeze loader work together to stack bales rapidly for carting.

bale carriers means that the system can be operated by one man up to the stacking stage. Because the bales are stacked in blocks in the field, weather exposure at this stage is minimised.

But the major economies in time are derived from the ability to stack bales mechanically in the barn. Hand-stacking requires 19 man-minutes/tonne compared to 6 man-minutes/tonne when bales are stacked with a tractor loader.

Flat 8 or flat 10 accumulators can be used in conjunction with an on-floor grain dryer to give a simple system of barn drying which does not involve a high time-cost in double-handling the bales. This is particularly so if the bales in question are large and rectangular rather than small.

Loose hay

We are now seeing the beginning of a trend which, if it continues, may signal a 'merger' between hay and silage. The trend is away from the bale and towards the harvesting and storage of loose hay. The stored crop is removed for feeding by a high-reach front-end grab loader.

Forage wagons are the most common form of harvesting equipment, but at Boxworth EHF a metered-chop forage harvester has been used to produce a chopped hay suitable for inclusion in a complete diet mixer-wagon. The attraction of the forage wagon is that it can be the basis of a one-man system in which the hay is placed in a dump box and then blown into the store.

Loose hay can be dried in the store, using pallets mounted on sleepers to form the drying floor. Provided the quantity harvested each day is not too great, wall pressures are relatively low and hay can be harvested at up to 50 per cent moisture content. The crop is blown continuously with ambient air during loading, for as long as is necessary to reduce the moisture content to the point at which no heating takes place overnight.

Barn-drying

Despite the fact that barn-drying allows hay to be removed from exposure to the weather in the field at an earlier time than with field-cured hay, it has remained a specialist activity in the UK. The process involves additional capital charges, as well as higher labour costs, and the fixed equipment may be under-utilised if the weather is good or if crop yield is low.

The additional costs of barn-drying may not always be recouped, therefore, in reduced losses and improved feed value. There may also be problems in matching the capacity of the dryer to the bale handling system of the farm. Surveys have shown that most farmers who had decided not to adopt barn-drying had done so because they regarded a dryer as either too expensive or unnecessary.

Barn-drying has been used successfully for many years at Drayton EHF. The capacity of the barn-dryer is 120 tonnes. This allowed 240 tonnes of hay to be dried each year — equivalent to 1000 tonnes of silage. The hay was made from grass which received 100 kg N/hectare and which was grazed until mid-March. The fields were then closed up for six weeks before being cut. The barn-dried hay was a high-value product of 70 D-value.

Success in barn-drying is achieved through being flexible; only cut enough at the start to barn-dry. If the weather stays good, cut some more and earmark the first batch for baling with a preservative. If the weather continues to stay good, field-dry the first batch, leaving the option open to either treat or barn-dry the remaining batches. In practice, this means that the capacity of the dryer need only be about a third of the total tonnage of hay.

Significantly, losses in the field and in store can be as low as 15 per cent with this flexible approach.

On mixed arable and livestock farms the dryer can be made more cost-effective if it doubles as a grain dryer at harvest time.

References

1. Zimmer, E. (1977) *Proceedings of an International Meeting on Annual Production from Temperate Grassland*, Dublin, 121—125.
2. Klinner, W.E. (1982) paper presented at the John Deere Grassland Seminar, Dublin.
3. Redman, P.L. (1972) *A Study of Bale Handling Methods*, report of an ADAS survey by Farm Mechanisation Advisory Officers.

7 Upgrading Low-quality Crops

The technology of upgrading was developed originally for straws. It is now being applied to grasses and whole-crop cereals, and the opportunities are considerable. Consider a year like 1983. Continuous rain for the month of May was followed by a long period of dry weather in June and July. First-cut silage was either a wet scramble between rain showers, or too late for crops of high digestibility to be harvested. Haytime, by contrast, was dry. But again the crops were often of low feed value because of their advanced stage of growth at cutting.

Several months elapse between storing the conserved crop and removing it from the store for feeding. There is therefore the prospect of treating it at harvest so that by the time the animal receives the product, the feed value is higher than at the start of the storage period.

A further bonus is that delayed harvest of grass crops is reflected in a higher yield of dry matter (DM) per hectare of land. The objective is to combine this high yield with reasonably high feed value, by upgrading.

Sodium hydroxide

Sodium hydroxide has been used for many years as the cheapest and most effective chemical for increasing the digestibility of low quality forages and straws. The alkali breaks the bonds between lignin and cellulose, thus increasing the potential digestibility of the cell wall or fibrous fraction of the crop. There is also a swelling of the fibre, which increases its accessibility to the enzymes secreted by the cellulose-digesting bacteria in the rumen. Thus not only the *extent* of digestion (digestibility) but also its rate is increased.

But there are problems associated with the use of sodium hydroxide. First, the chemical is difficult to handle on farms and extreme care has to be taken to ensure that it does not come into contact with the eyes or the skin. Secondly, the relatively high concentration of sodium in the treated crop is excreted by the animal via the urine. This means that urine output is increased, particularly if the alkali-treated material contributes significantly to the diet.

The response in digestibility of hay to addition of sodium hydroxide at 4 per cent of the crop dry matter (DM) depends on stage of maturity: the older the crop, the greater the response. In the trial shown in fig. 7.1, which was carried out at Reading University,[1] a three-week delay in harvesting resulted in a 10-unit decrease in digestibility. Treatment of the more mature hay crops gave improvements in digestibility so that they equalled untreated hay made three weeks earlier, from lower-yielding crops.

Treatment of straws with sodium hydroxide at 4 to 5 per cent of the crop DM may be expected to result in an increase in metabolisable energy (ME) content of the stored product of 1.5 MJ per kg DM (15 units increase in D-value), providing mixing of alkali and feed is adequate and uniform.

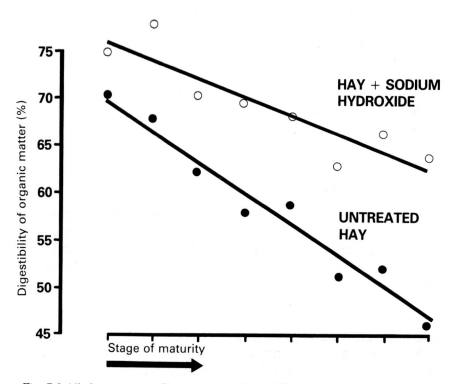

Fig. 7.1 Alkali-treatment of hay increases digestibility.

A final problem with the use of sodium hydroxide is that there is an increased demand for nitrogen by the rumen microbial population, because of the improvement in digestibility. This means that additional nitrogen is needed, especially in the case of straws and very mature crops of grass, in order to realise the improved digestion characteristics of the treated product.

Ammonia

Ammonia has now largely superseded sodium hydroxide as the preferred chemical for upgrading because it not only acts as an alkali in the same way as sodium hydroxide, it also contains supplemental nitrogen. But ammonia poses greater handling problems than sodium

Plate 7.1 Ammonia-treatment of straw: large round bales are placed in an 'oven' prior to treatment with anhydrous ammonia and heat.

hydroxide because it is more volatile. For this reason it is an operation for the specialist contractor rather than the farmer.

Two techniques are in use on farms, one involving a 24-hour 'oven' process in which bales are placed in a box and treated with anhydrous ammonia and heat. Unreacted ammonia is flushed out with air towards the end of the treatment period.

The other process comprises making a stack of bales which is sealed by top and bottom sheets of polythene. Ammonia (either anhydrous or aqueous) is then injected into the stack. A variation on this technique has recently been introduced whereby large round bales are injected with anhydrous ammonia as they are put into a 'sausage' of polythene. Injection occurs whilst the bales are on the fore-end loader of the tractor, which also carries a small tank of ammonia.

Plate 7.2 Injecting large bales with anhydrous ammonia as they are inserted into plastic 'sausages'.

Ammonia has theoretical advantages over sodium hydroxide. Apart from increasing the content of nitrogen in the treated crop, there is no excess·of sodium in the diet. Trials with hay at the Grassland Research Institute have shown that the stack method gave consistently higher values for digestibility compared to the oven method, possibly because more of the ammonia was bound to the plant cells in the case of the oven treatment. But farm trials with straws of lower initial quality have shown that more consistent upgrading is obtained by using the oven method.

Plate 7.3 Injecting a stack of bales, sealed in polythene sheeting, with aqueous ammonia.

The recommended rate of addition of ammonia is 35 kg NH_3 per tonne of dry matter. At this level of addition, the increase in crude protein equivalent (nitrogen x 6.25) due to bound nitrogen is likely to be about 8 percentage units. Treated straws and mature hay crops would therefore be expected to analyse at about 11 per cent and 17 per cent crude protein, respectively.

In a trial in Northern Ireland,[2] baled hay was purchased and treated with anhydrous ammonia by the stack method. When finishing beef cattle were given either untreated or treated hay as the sole feed, there was a marked response in daily gain to hay that had been treated with ammonia (see table 7.1).

Table 7.1 Response by beef cattle to hay treated with anhydrous ammonia.

	Untreated hay	Ammonia-treated hay*
Digestibility of organic matter (%)	53	66
Liveweight gain (kg/day)	0.49	0.79

* 30 kg NH_3 per tonne DM.

Similarly, in a Danish trial (see table 7.2),[3] dairy cows were given the same hay crop, which was of quite high initial quality (65 D-value), either as untreated barn-dried material, or as ammoniated moist hay, treated by the stack method. The main feature of this trial was that despite the high quality of the barn-dried hay, the cows ate 20 per cent more of the ammonia-treated material. They gave slightly more milk, and gained more weight on the ammonia-treated hay than on the barn-dried crop.

The advantage of the oven method is that the treatment process is completed in 24 hours, whilst with the stack method Scandinavian trials have shown the best responses to be achieved if the stack is kept sealed for two months before opening.

Table 7.2 Response by dairy cows to hay treated with anhydrous ammonia.

	Untreated hay (barn-dried)	Ammonia-treated hay* (70% dry matter)
Intake of dry matter (kg/day)		
hay	8.1	9.7
concentrate	6.3	6.3
Milk yield (kg/day)		
(4% fat-corrected)	15.0	15.8
Weight gain (kg/day)	0.28	0.76

* 35 kg NH_3 per tonne DM.

It is essential to have uniform distribution of the ammonia through the crop when the stack method is adopted. To achieve this it is advisable to avoid using either very dry or very moist material. Ideally the crop moisture content, for addition of anhydrous ammonia, should be about 30 to 35 per cent. It is also advisable to aerate the stack for two to three days prior to feeding.

Urea

The problem of the safe handling of ammonia on farms may be overcome by using urea instead. The object here is to generate ammonia within the crop during the storage period by hydrolysis of the urea. This is brought about by enzyme (urease) activity. Urease is a naturally-occurring enzyme, and trials at both the Grassland and Rowett Research Institutes, and elsewhere in Europe, have demonstrated that there is extensive hydrolysis of urea to ammonia following its addition to hays, straws and whole-crop cereals.

A further benefit from the use of urea is that it appears to act as a very effective preservative for moist hay (see table 7.3). Perennial ryegrass was harvested[4] as 'very moist' hay (57 per cent DM) and stored for 120 days in woven polypropylene sacks which allowed air to pass through the crop. The level of addition of urea was chosen to approach the equivalent of 3 per cent of ammonia, and during the first month of storage at ambient temperature almost all the urea was converted to ammonia with most of the conversion occurring in the first week.

The untreated hay, which had a D-value of 54 per cent at harvest, deteriorated badly during storage so that at the end of the period almost 20 per cent of the digestible organic matter had been lost and the percentage of digestible organic matter in the remaining dry matter (D-value) had decreased by 3 percentage units (see table 7.3).

Table 7.3   Preserving and upgrading moist hay with urea added at harvest.

	At harvest	After 120 days storage in air	
		Untreated	Urea-treated (6% of DM)
Loss of digestible organic matter (%)	—	19	4
D-value *in vitro* (%)	54	51	58

By contrast, there was very little loss in the case of the urea-treated crop, which showed an improvement of 4 percentage units in D-value during storage, and an improvement of 7 percentage units over the stored, untreated material.

Crops with low contents of nitrogen such as mature hays and straws will show a response in digestibility to addition of urea at the time of feeding. This is because there is insufficient nitrogen in the plant material to maintain the optimal rate of microbial digestion in the rumen. Addition of urea at harvest, rather than at feeding, not only provides the supplemental nitrogen required by the rumen microbial population; it also utilises the time which elapses between harvesting and feeding to 'pre-digest' a proportion of the feed.

Alkali treatment of whole-crop cereals

If alkali-treatment works with straws and hay crops, why not with whole-crop cereals? The attraction of making silage from the whole crop of wheat or barley is that there is only one harvest, compared to two or more in the case of conventional grass crops. The disadvantage is that at maximum yield, the overall energy value of the product is much lower than with grass, which is usually cut at a less-mature stage of growth.

A recent trial has been completed at the Grassland Research Institute in which Hereford x steers, initially 12 months old, were finished on whole-crop winter wheat silage-treated, as before, with 6 per cent of urea at harvest. The crop, Huntsman, was harvested two weeks prior to combine-ripeness and yielded 10.7 tonnes of dry-matter per hectare. The performance of the cattle is shown in table 7.4.

Table 7.4 Treated whole-crop winter wheat* for finishing beef cattle.

	Low concentrate (0.8 kg/day)		High concentrate (2.8 kg/day)	
	− urea	+ urea	− urea	+ urea†
Liveweight gain				
(kg/day)	0.71	0.85	0.93	1.27
(kg/ha)	1460	1723	2304	2873

* Harvested two weeks before combine-ripeness at 60% DM.
† Urea added at harvest at 6% of crop DM.

Weight gains were improved at both low and high levels of concentrate supplementation, yet intake of the treated and untreated silages was similar at both levels of supplementary feeding. This suggests that addition of urea at the time of ensiling was associated with improved digestibility. The output of liveweight gain per hectare (table 7.4), at over 2800 kg for urea-treated whole-crop wheat silage, is comparable to the levels obtained from high-quality grass silage made from three cuts during the season and given to beef cattle in the grass silage beef system. This system is currently being studied at Rosemaund Experimental Farm and the National Agricultural Centre Beef Unit, where gross margins of over £1000 per hectare have been recorded.

Is upgrading economic?

The alternative techniques are summarised in table 7.5. Recommended rates of addition of additive, together with estimated treatment costs per tonne of crop dry-matter are shown in table 7.6. Ammonia is likely to be more expensive than either sodium hydroxide or urea. In the case of urea, proprietary solutions which also contain essential minerals are available.

As yet there is insufficient information available to judge the probable animal responses to urea, but in the case of both sodium hydroxide and ammonia very substantial increases in intake of meta-

Table 7.5   Upgrading conserved forages: alternative techniques.

	Sodium hydroxide	Ammonia		Urea
Process	To bales via straw processor	Oven method (small or	Stack method large bales)	To bales via straw processor
Additive	27% solution of sodium hydroxide in water	Anhydrous ammonia	33% solution of ammonia in water	30% solution of urea in water
Batch size	continuous process	0.9 tonnes	20 tonnes	continuous process
Treatment time	24 hours	24 hours	8 weeks	8 weeks
Temperature	ambient	90°C for 15 hours	ambient	ambient

Table 7.6 The Wilkinson guide to upgrading low-quality crops and by-products.

Chemical	Recommended rate of addition (kg/tonne DM)	Estimated cost (£/tonne DM)		Probable animal response	
		Chemical alone	Total	Intake of DM	Intake of ME
Sodium hydroxide	40	8	15 to 20	+ 50%	+ 80%
Ammonia	35	12	20 to 25		
Urea	60	10	12 to 15	insufficient information available	

bolisable energy have been recorded with a variety of low-quality crops and by-products.

The technique of upgrading therefore allows greater use to be made of unpalatable feeds in the diet of productive livestock. If untreated straw, for example, is valued at less than £20 to £25 per tonne 'as fed', and the alternative forage feed is hay costing over £50 per tonne delivered to the farm, then from what is known about the value of average hay and treated straw as feeds, upgrading is likely to be cost-effective, at least in the case of sodium hydroxide and ammonia.

Enzymes

Genetic engineering applied to agriculture introduces exciting future possibilities for the upgrading of low-quality crops and by-products. Hitherto, we have considered chemicals which, in the main, pose problems of safe handling on farms, and which, because they are alkaline, are not effective with wet crops. An alternative approach could involve the 'pre-digestion' of the plant cell wall fraction by the addition of selected cellulolytic and lignolytic enzymes. These enzymes might eventually be produced in large quantities under controlled conditions by specially engineered micro-organisms. Such enzymes would be particularly useful for use with wet silage crops, since the products of their action — sugars — might be available for fermentation to lactic acid in the silo. At present, though, there is no evidence that the addition of cellulase enzymes can match the large improvements in feed value following the addition of relatively large quantities of cheap alkali.

References

1. Mwakatundu, A.G.K. and Owen, E. (1974) *East African Forestry Journal,* **40**, 1–10.
2. Wylie, A.R.G., Department of Agriculture for Northern Ireland.
3. Winther, P. *et al.* (1983) *Report No. 9*, Danish Research Service for Soil and Plant Science, Copenhagen.
4. Tetlow, R.M. (1984) *Animal Feed Science and Technology.*

8 Winter Feeding

Previously, one was taught that there should always be a spare bay of hay left over in the barn at the end of the winter. Nowadays we should perhaps translate that to a spare 100 tonnes of silage left in the silo. This should be sealed at the end of winter so that it doesn't deteriorate during the summer months.

There are two good reasons for planning to have excess silage at the end of the winter. If you run out altogether in late winter, the chances are that a lot of other people have also used up their conserved forage and bulk feeds will be very costly to purchase at this time. In fact it is usually better to buy concentrates: they can be cheaper per megajoule (MJ) of metabolisable energy (ME) than hay at £70 or £80 a tonne.

Secondly, a reserve of silage or hay available through the summer

Plate 8.1 Feed budgeting is essential for the winter period. It is necessary to estimate as accurately as possible how much silage is in store.

allows you to give the animals a supplement when grazing conditions are poor. Excessively wet or dry weather causes shortages in the supply of available grass for grazing. A buffer feed at times like these can give useful responses in production.

Estimating the supply of silage

It is important to have a clear idea of how much silage is in the silo. Inaccurate estimation of density, for example, can cause large errors in assessing the quantity available for the winter period.

The best way to judge the quantity of silage in a silo is to take core samples and measure the density of the material in the core. If this is not possible, then the next best method is to estimate density from dry matter content.[1] The equations in table 8.1 relate to bunker silos filled to a depth of 2 metres. As the dry matter content of the silage increases, dry matter density is also increased, but bulk density decreases because compaction is lower with crops of higher dry matter content.

Budgeting for the winter period

Dairy cows

Feed budgeting for dairy cows comprises setting targets for milk production, for the total requirement of metabolisable energy (ME) and for the quantity of purchased concentrates or compounds. Then the amount of silage or hay required to meet the remainder of the

Table 8.1 Estimating density of grass silage in bunker silos.

Density of dry matter (kg/m^3) $= 65 + 4.0 \times DM\%$

Density of fresh weight (bulk density) (kg/m^3) $= \dfrac{6500}{DM\%} + 400$

Examples

Dry matter (DM) (%)	Density (kg/m^3)	
	Dry matter	Fresh weight
18	137	760
22	153	695
26	169	650
30	185	615

total ME requirement is calculated. At this stage it is necessary to check that the cow is likely to be able to eat the diet which is proposed, that the supply of minerals and vitamins in the diet is adequate, and that the total amount of silage or hay required for the winter period does not exceed the supply.

The steps are outlined in table 8.2 with an example for a cow weighing 600 kg liveweight with a target milk yield over a 6-month

Table 8.2 Feed budgeting for winter milk.

	Example
Step 1 Set target milk yield and days on winter feed.	4000 litres 180 days on winter feed
Step 2 Calculate total requirement for ME in gigajoules (GJ).	A 600 kg liveweight cow will require: ME for maintenance 12.5 ME for milk (5.3 GJ/1000 litres) 21.2 Total $\overline{33.7\,GJ}$
Step 3 Set target concentrate use.	1.5 tonnes fresh weight 1.33 tonnes dry matter (DM)
Step 4 Calculate ME from concentrate and ME required from silage or hay.	ME content of concentrate (from analysis) 1.33×12.0 GJ/tonne DM = 16.0 GJ ME from silage or hay $33.7 - 16.0$ = 17.7 GJ
Step 5 Calculate total silage or hay DM required	ME content of silage or hay (from analysis) = 10.2 GJ/tonne DM Total DM required = $17.7 \div 10.2$ = 1.74 t
Step 6 Check cow is likely to eat diet. Check supply of minerals and vitamins is adequate	Total daily intake of DM (kg) = $0.022 \times$ liveweight + $0.2 \times$ milk yield per day Daily yield = 22.5 litres Intake = $0.022 \times 600 + 0.2 \times 22.5$ = 17.7 kg Proposed diet: 1330 kg concentrate DM + 1740 kg forage DM = $\overline{3070} \div 180$ days = 17.1 kg

Note: If cow cannot eat diet, or if quantity of silage or hay exceeds supply, increase concentrates.

* A gigajoule (GJ) is a thousand megajoules (MJ).

winter period of 4000 litres. The procedure may be varied if, for example, supplies of silage or hay are known to be low. Step 3 then becomes a calculation of the total supply of conserved forage dry matter per cow for the winter period. Step 4 is the same as before, but Step 5 becomes the calculation of the total concentrates required for the winter.

It is vital to know the ME content of the silage before the winter period commences. If silage quality is low, the cow may not be able to eat enough to meet her total energy requirements. In addition, the quality of the silage has a marked effect on the total amount of concentrate required, as illustrated in table 8.3. Here, three silage qualities are assumed for a farm where 720 tonnes of settled silage is available for the winter.

In the case of the lowest quality silage, the cows are unlikely to eat their full daily allowance of silage; milk yield will fall as a consequence. It is therefore necessary to reduce the daily allowance and make up the shortfall with additional concentrates. The difference between the highest quality and the lowest quality silage is 0.4 tonnes concentrate per cow.

Table 8.3 Effect of silage quality on feed budgets for winter milk.

100-cow herd, 720 tonnes settled silage of 25% dry matter.
Estimated allowance for 180-day winter, 10 k DM/cow/day.
Total milk 4000 litres/cow; total ME required = 33.7 GJ or 187 MJ/day (from table 8.2).

	ME of silage (MJ/kg DM)		
	9.3	10.0	10.7
DM from silage (kg/day)	10.0 (9.0)	10.0	10.0
ME from silage (MJ/day)	93 (83.7)	100	107
ME required from concentrates (MJ/day)	94 (103.3)	87	80
DM from concentrates (kg/day)	7.8 (8.6)	7.3	6.7
Total DM	17.8 (17.6)	17.3	16.7
Will cow eat diet?	(Total intake (table 8.2) = 17.7 kg/day)		
	No	Yes	Yes

In the case of the lowest quality silage, the allowance is too high and must be reduced to 9 kg DM/day (figures in brackets).

| Total concentrate (t/cow) | (1.72) | 1.45 | 1.33 |

Table 8.4   Winter finishing of beef cattle: alternative feed strategies and feed budgets.

Daily feed			
Rolled barley (kg)	2.4	2.8	3.2
Silage (kg)*	21	20	19
Performance†			
Daily gain (kg)	0.7	0.8	0.9
Finishing period (days)	285	220	165
Slaughter weight (kg)	525	500	475
Feed budget			
Rolled barley (t)	0.67	0.62	0.53
Silage (t)	6.0	4.4	3.2

* 9.5 MJ ME/kg DM, 25% DM content fed to appetite
† Friesian steers

Beef

Beef cattle gains can be varied quite substantially by altering the ratio of conserved forage to concentrate in the diet. In the case of winter finishing, this results in different feeding periods because faster-grown cattle are ready for slaughter earlier than slower-grown animals, though at a lighter weight. A slightly higher level of concentrates can therefore be used to shorten the finishing period. Overall, there is little change in total concentrate requirements but a large reduction in the quantity of silage needed (table 8.4).

Sheep

Feed budgeting for the winter feeding of sheep should take into account the bodyweight, condition and age of the ewe, and whether or not she is carrying a single lamb or twins. The critical periods are the first month of pregnancy when it is necessary to avoid sudden, short-term reductions in energy intake which might lead to embryonic loss. In late pregnancy the aim should be to prevent excessive depletion of the ewe's body reserves which will lead to reduced lamb birthweight, less vigour in the lambs, and a delayed onset of lactation in the ewe.

Guidelines for the feeding of ewes of two different weights during late pregnancy are given in table 8.5. They are from the Meat and

Table 8.5   Concentrates and hay for pregnant ewes (kg/day).

Weight of ewe (kg)	Foetuses	Weeks before lambing					
		6		4		2	
		Hay	Concentrate	Hay	Concentrate	Hay	Concentrate
50	Single	0.9	0.2	0.9	0.3	0.9	0.4
	Twins	0.9	0.3	0.9	0.4	0.9	0.5
70	Single	1.3	0.2	1.3	0.3	1.3	0.5
	Twins	1.3	0.4	1.3	0.5	1.3	0.7

Livestock Commission's guide *Feeding the Ewe*. The figures are based on hay of medium quality and on 18 per cent crude protein concentrate mix containing 15 per cent soyabean meal and 85 per cent mineralised barley.

Feed budgets for ewes of two different weights, carrying single or twin lambs, for the complete winter period are given in table 8.6. In addition to the total concentrates required during the winter, some supplementary feed is usually provided in early spring when the flock is at pasture.

How much silage or hay will animals eat?

Appetite is of major importance in achieving high levels of production from diets based on conserved forage, yet we know surprisingly little about the factors affecting the intake of different silages and hays, particularly when given to dairy cows. Perhaps this is because few researchers have been prepared to give silage or hay as the sole

Table 8.6   Winter feed budgets for ewes.

Foetuses	Weight of ewe:			
	50 kg		70 kg	
	Single	Twins	Single	Twins
Concentrate (kg/ewe)	20	25	25	32
Silage or hay (kg DM/ewe)	150	150	200	200

feed to lactating cows, and also because several factors can interact to influence intake. We need to be able to predict the intake of a silage or a hay from its chemical analysis, and then to estimate the probable intake when given in a mixed diet.

Dr Malcolm Castle at the Hannah Institute fed his Ayrshire cows (470 kg liveweight) on silage as the sole winter feed for four years, between 1973 and 1977.[2] During this period he measured intake and milk production from eight high-quality, well-preserved silages. His results are summarised in table 8.7.

Table 8.7 Silage as the sole feed for dairy cows*.

Analysis of silage	
Dry matter (%)	27
pH	3.96
ME (MJ/kg DM)	11.3
Intake of silage (kg)	11.3
(kg/100 kg liveweight	2.41
Yield (kg a day)	
Milk	14.4
Fat	0.61
Protein	0.44
Lactose	0.66

* Average of eight trials

Intake is the ultimate limitation on production. If, as with mixed diets of hay and concentrates, intakes of 2.9 to 3.0 per cent of live-weight could be achieved, then daily yields of 20 kg of milk could theoretically be obtained from diets of silage alone. Intake is probably limited by the bulk or 'fill' in the rumen of the animal. Thus the heavier (bigger) the animal the more forage it can physically accommodate at any one time in its digestive tract. Castle's research indicates that silages of high quality are eaten at levels of dry-matter consumption of about 2.4 per cent of liveweight. This value has been confirmed in pilot studies at the ICI research station at Jeallots Hill, Berkshire, where Friesians (590 kg liveweight) were given silage of 11.2 MJ ME per kg DM as the sole feed. The cows ate 13.6 kg dry matter per day, or 2.3 per cent of liveweight. Peak yield was 25 kg per day and the cows averaged 21.4 kg milk over a 106-day feeding period from calving to turnout to pasture.

One way round the problem of production being limited by intake, therefore, is to go for heavier cows. Many dairy farmers have, of course, already done this by switching to larger-framed Friesians or to Holsteins.

Probable levels of intake of conserved forage of high quality when given to dairy cows, beef cattle or sheep as the sole feed are given in table 8.8. Whilst these values may be useful for feed budgeting, they really serve as a baseline from which to assess the likely intake of more moderate quality forages, and also the extent to which intake may be reduced when concentrates are included in the diet.

Factors affecting intake of conserved forages

Preservation quality has an over-riding influence on the intake of silages, followed by energy content or stage of crop maturity at harvest, and length of chopping. In the case of hays, stage of maturity exerts the major influence on intake. The changes in intake due to these factors, relative to those of a high-quality forage with an ME content in excess of 10.5 MJ/kg DM, are summarised in table 8.9.

Table 8.8   Potential intake of conserved forage*.

	Liveweight (kg)	Intake of DM (kg/day)
Dairy cows	500	11.5
	550	12.5
	600	13.5
Beef cattle	200	5.0
	300	6.5
	400	8.0
	500	9.0
Sheep	30	0.7
	40	1.0
	50	1.2
	60	1.4
	70	1.6
	80	1.8

* Well-preserved forage with an ME content over 10.5 MJ/kg DM given as the sole feed.

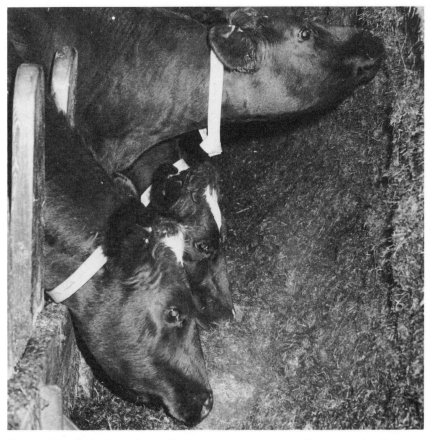

Plate 8.2 Intake is the ultimate limitation on production. When self- feeding, allow at least 20 cm of silage face per cow if intake is not to be restricted by inadequate access.

The effects of energy content and length of chop on silage intake are reduced when preservation quality is poor. The effect of chop length is likely to be greater with silage of low ME value.

A silage of 9.0 MJ ME, of moderate preservation quality, made with a flail harvester is likely to be eaten at only 65 per cent of the level of a well-preserved short-chopped silage of high ME content.

The above discussion relates to *ad lib* feeding, that is, silage available to the animals at all times. To achieve true *ad lib* intake, not only must there be silage available, but sufficient space should be given to ensure that each animal can spend sufficient time at the silo face or at the trough. Recommended allowances are:

(a) trough-feeding: at least 15 cm of trough space per cow;

Plate 8.3 Digestibility is a major influence on the intake of conserved forages, especially of hay and straws.

(b) self-feeding: at least 20 cm of face per cow, with the height of the face less than 2 metres.

Supplements

Supplementary feeds, especially concentrates, fulfil an essential role of enriching the diet with energy and also providing undegraded dietary protein (UDP) to enhance the supply of amino acids for tissue synthesis or milk production. Another important role is that of 'buying in land', that is, of increasing the total winter feed supply to the stock.

When animals are given concentrates, they usually respond by eating less silage or hay. This *substitution* of one feed by another is important because it means that the response in milk yield or weight gain is lower than would be predicted from simply adding together the quantities of ME supplied by the different feeds. Substitution rates vary with the quality of the forage and with the level of input of concentrate. They can also vary between different types of supplement.

Under most feeding regimes the substitution rate is probably about 0.5 kg decrease in forage DM intake per kg increase in concentrate DM intake. This effect can be beneficial, as we saw earlier,

Table 8.9   Factors affecting intake of silage and hay.

	Relative intake
Metabolisable energy (MJ/kg DM)	(well-preserved silages and hays)*
over 10.5	100
10.0–10.5	95
9.5–10.0	90
9.0– 9.5	80
Preservation quality (silage) Ammonia nitrogen (NH$_3$-N) as percentage of total nitrogen	
Good (less than 10%)	100
Moderate (10–15%)	95
Poor (more than 15%)	90
Chop length (silage)	
Short (less than 10 m)	100
Medium (10–50 mm)	95
Long (more than 50 mm)	85

* With poorly-preserved silages there is relatively little change in intake with change in ME content.

in reducing the feeding period of finishing beef cattle and the total requirement for conserved forage. With dairy cows given high-quality, well-preserved silage *ad lib*, however, the substitution rate may be much greater than 0.5 so that a response in milk yield is barely noticeable.

Different supplements appear to give different rates of substitution – at least with silage. Dr Castle's experiments have included a wide range of contrasting supplements (table 8.10), though they have been given in different quantities so it is important to distinguish between the effect of *type* of supplement and *level* of supplement on substitution rate.

Long hay, regardless of its ME content, is of little value as a silage supplement. This means that hay is of most value when silage is in short supply. Dr Castle considers that a useful supplement for silage would be a concentrate mixture of high protein content and with a low barley content. Such a supplement would presumably be relatively high in ME (over 12.5 MJ/kg DM) and also high in UDP.

Table 8.10   Substitution rates of different supplements.

	Substitution rate*
Hay	0.8
Barley	0.5
Sugar-beet pulp	0.4
Dried grass cubes	0.4
Barley + protein (e.g. soya)	0.3
Soyabean meal	0.0

* kg decrease in silage DM intake per kg increase in supplement DM intake.

It could then be given at a low rate per kg milk to achieve a low rate of substitution. This type of feed would be particularly valuable when silage is in abundant supply and of high quality.

To achieve low rates of substitution, several points should be borne in mind:

(a) choose a supplement relatively high in protein or digestible fibre;
(b) feed the supplement frequently, in small amounts per feed;
(c) choose dry rather than wet supplements;
(d) choose a supplement of high palatability (sweet);
(e) correct any nutrient deficiencies in the forage (e.g. low nitrogen in maize silage).

Sodium hydroxide-treated straw, because of its high pH and low nitrogen content, may be a very useful complementary feed to grass silage, which has a low pH and a relatively high content of degradable nitrogen. Molasses may also offer value for money, as a means of increasing palatability and of providing extra energy in the diet. With the rapid expansion of the oilseed rape crop in the UK, there may eventually be scope for using processed, detoxified whole rapeseed. The product is likely to have both a high ME and high protein content. Finally, there is considerable scope for making better use of abattoir wastes, such as fat and blood, as supplements particularly if they can be processed so that palatability problems are overcome and protection against rumen degradation is also achieved.

References

1. Smith, M.S. (1980) *Technical Note NC/80/2*, ADAS.
2. Castle, M.E. (1982) *Silage for Milk Production*, Technical Bulletin No. 2, Hannah Research Institute, 127—150.

9 Profitable Milk from Grass

'The most profitable cow', said Bobby Boutflour, 'is the one with the highest yield coupled with the greatest margin between cost of production and selling price'. Thirty years ago, when our knowledge of dairy cow nutrition was rudimentary, men like Boutflour were achieving single lactation yields of 9000 litres from their best cows — yields which were comparable to the *lifetime* performance of the average cow.

That was before the dairy industry of Europe, with its small family farms, learned how to achieve higher yields without incurring excessively high costs of production. Now, with over-production of milk the biggest single headache of the administrators of the Common Agricultural Policy, the economic climate is rather different to that of the fifties.

European dairying compared

The lastest analysis of European dairying, by Mr Steve Amies of the Milk Marketing Board,[1] highlights the importance of variations in selling price, inflation and interest rates in affecting profit. Since Boutflour's time, these factors, which are largely outside the individual farmer's control, have increased in importance.

If milk prices in the European Community were reduced to levels currently prevailing on the world market, only the Irish dairy industry could survive relatively intact. In Ireland, milk is produced from grass with relatively low levels of concentrates and nitrogen. Farm gross margins between 1978/9 and 1982/3 were lower than in UK because output of milk was also relatively low. But because the average size of dairy farms is only half that in UK, overhead costs were much lower. In addition, the Irish farmer has been cautious in borrowing money. This attitude probably reflects the fact that historically the selling price of milk in that country has been lower than elsewhere in Europe. The consequence is that although gross margins were lower in Ireland, profit per cow was almost 50 per cent higher than in UK (table 9.1).

Table 9.1   Costs of milk production in UK and Ireland, 1978/9 to 1982/3.

	UK	Ireland
Herd size	130	64
Farm size (ha)	81	48
Milk yield (litres/cow)	5298	3511
Stocking rate (cows/ha)	1.95	1.75
Inputs		
Concentrates (t/cow)	1.77	0.57
Nitrogen (kg/ha)	249	180
Farm gross margin (£/cow)	345	261
Overhead costs (£/cow)		
Paid labour	58	10
Buildings, machinery	97	29
Money	42	22
Profit (£/cow)	121	179

Costed herds in UK were bigger, had higher levels of inputs and generated higher farm gross margins than those in Ireland. But overheads were relatively greater and profit per cow in Irish herds was 48 per cent higher.

During the five-year period reviewed in the MMB report, profits were reduced by the combined effects of increased costs and a fall in real terms in the selling price of milk. The report highlights the better 'survivability' of the smaller family farm under such adverse economic conditions. On the larger farms paid labour is a cost that has to be met irrespective of whether or not the farm is profitable. On the other hand, by employing paid labour dairy farmers in the UK should have greater opportunities for expanding their businesses, provided they remain profitable, perhaps by diversification into milk processing or retailing.

It would be wrong to conclude that the Irish model is appropriate for the whole of the UK. We should perhaps look also to Holland, where milk yield per cow averages almost 5500 litres from 1.5 tonnes of concentrates, and where stocking rates approach 3 cows per hectare.

Comparison of milk yields in seven different parts of Europe (see fig. 9.1) shows that the farms costed by the MMB have a relatively

Yield/cow (tonnes)

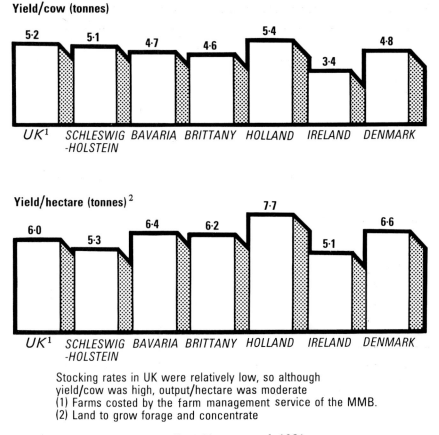

Stocking rates in UK were relatively low, so although
yield/cow was high, output/hectare was moderate
(1) Farms costed by the farm management service of the MMB.
(2) Land to grow forage and concentrate

Fig. 9.1 European dairying, milk yields compared, 1981.

high average yield per cow. But when the total land area used to
grow forage and cereals for the concentrates is considered, our
performance is only moderate. Concentrates are used in Holland and
Denmark to boost stocking rate on the smaller farms with limited
land. High-yielding grasses and legumes, such as maize and lucerne,
are exploited because of their suitability to the drier climate. The use
of fodder beet in Denmark, with its very high yield of metabolisable
energy (ME) per hectare, has enabled stocking rates there to be
pushed over 3.2 cows per hectare, whilst in the UK, the top 25 per
cent of FMS-costed farms averaged a stocking rate of only 2.5 cows
per hectare in 1981/82. There is clearly considerable scope here for
improvement in the way we integrate grass with other forage and
concentrate feeds to achieve high stocking rates.

The extent to which the Dutch and the Danes are ahead of

farmers in the UK in exploiting high-yielding forages is well illustrated in table 9.2. The *average* level of utilised metabolisable energy output in Holland, at 102 GJ per hectare, is 55 per cent higher than the average of FMS farms in the UK, and equal to the output achieved by our top grassland farmers (see table 9.3). This high level of

Table 9.2 Utilised metabolisable energy from forage (GJ/ha).

	1981
Holland	102
Denmark	92
Bavaria	77
Brittany	75
UK	66
Ireland	64
Schleswig-Holstein	57

Despite an average use of 250 kg N per hectare on FMS farms, UME output from forage barely surpassed that achieved in Ireland.

Table 9.3 Milk from grass: performance of top grassland milk producers, 1980/1.

Cows in herd	125
Milk yield	
(litres/cow)	5946
Concentrates	
(kg/cow)	1446
(kg/litre)	0.24
Nitrogen	
(kg/ha)	338
Stocking rate	
(cows/ha)	2.36
UME	
(GJ/ha)	104

Top grassland dairy farmers produced almost 6000 litres of milk from slightly less than 1.5 tonnes of concentrate. They used relatively high levels of nitrogen fertiliser and achieved a high UME output from grass.

technical efficiency is reflected in high profits — at least so far as Holland and Bavaria are concerned (see fig. 9.2). In Denmark, the high level of borrowing by dairy farmers has reduced profit per cow and per hectare to low levels.

The potential for profitable milk from grass

Rex Paterson would have had some very definite views on how to tackle the problem of over-production of milk in the European Community. In his own lifetime he was concerned that increased milk production would reduce the average milk price, and he preferred to see increased returns coming from increased stocking rate.

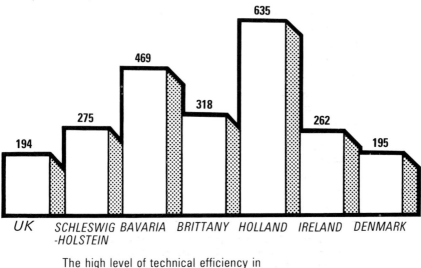

The high level of technical efficiency in
Holland was reflected in the profits achieved.

Fig. 9.2 European dairying, profits compared, 1978/9 to 1982/3.

He may well have been an advocate of a quota system for milk producers.

The Rex Paterson Memorial Study[2] illustrated the potential for high efficiency of milk production from grass. The average level of physical performance of the thirty-four herds in the study is shown in table 9.3. Milk yield was well above average, although concentrate input was below average. Top grassland farmers spent more money than the average on nitrogen, and in consequence their stocking rates were higher than average. Margin over feed and forage costs averaged £1247 per hectare in 1980/81, 84 per cent higher than the average for MMB-recorded farms in the same year.

Although yields were higher at higher levels of concentrate input, margins per cow and per hectare did not show a clear relationship to concentrate input (see table 9.4). Possibly this was because concentrates were replacing grass in the diet so that the marginal response in milk yield equalled the extra cost of the concentrate. The implication is that the actual rate of response was perhaps as low as 1 litre for each extra kilogram fed, which in financial terms is like exchanging £1 for 100 pence — there is no profit.

The balance between nitrogen, concentrate and stocking rate

A balance has to be struck between spending money on concentrates and spending money on fertiliser. The farmers in the Rex Paterson Memorial Study who chose to spend relatively more on concentrates, tended to spend less on nitrogen fertiliser, possibly because they considered the response to extra nitrogen might be uneconomic.

Table 9.4 Concentrate use, stocking rate, and margins in top grassland herds.

Concentrates (tonnes/cow)	Stocking rate (cow/ha)	Margin over feed costs	
		(£/cow)	(£/ha)
Below 0.9	2.22	610	1182
0.9–1.2	2.27	593	1170
1.2–1.4	2.49	600	1332
1.4–1.6	2.24	597	1213
1.6–1.8	2.34	592	1241
Above 1.8	2.52	636	1458

Differences in margin associated with different levels of concentrate use were small.

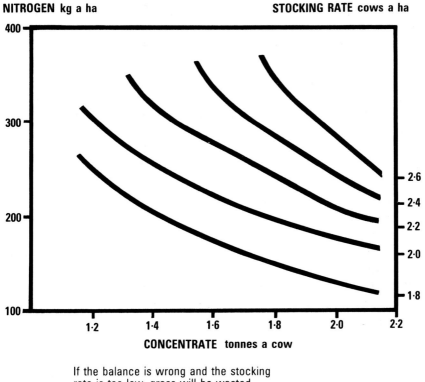

If the balance is wrong and the stocking
rate is too low, grass will be wasted.
If stocking rate is too high, milk yield will fall.

Fig. 9.3 The balance between nitrogen, concentrate, and stocking rate

The balance between concentrates and nitrogen fertiliser has also to be made in relation to stocking rate. Too high a level of either concentrate feeding or nitrogen use in relation to stocking rate will mean that grass will be wasted. Too high a stocking rate will lead to inadequate grass available for grazing and for conservation, and milk yields will fall. The actual levels of concentrate, nitrogen and stocking rate will depend on the potential of the cows in the herd to give milk, the potential of the site to yield grass, and the size of the herd in relation to the size of the farm. Smaller farms with limited land tend to have relatively high levels of input and high stocking rates.

Targets

The relationships between target stocking rate, concentrate use and nitrogen fertiliser are shown in fig. 9.3 for an autumn-calving herd yielding 6000 litres, and given silage of 10.2 MJ ME/kg dry matter.

They were derived by the ICI/GRI team which produced *Milk From Grass.*[3]

Ideally, a system of milk production designed to produce 6000 litres of milk per cow from 1 tonne of concentrates is a suitable target for profitable milk from grass.

Quotas have effectively placed a ceiling on farm output. This change in the economic climate focuses attention on producing milk to the quota *at the lowest possible input cost.*

The ICI/GRI team have spelt out the targets as they relate to average grass-growing conditions, and an autumn calving herd (see table 9.5). There is now a need to demonstrate that these targets are achievable in practice. They imply a reduction in stocking rate and an increase in the proportion of grass in the diet, compared to similar herds in the Rex Paterson Study. In that study, the autumn calving herds which used less than 1 tonne of concentrates per cow averaged less than 5500 litres of milk, used over 400 kg nitrogen per hectare and had an average stocking rate of 2.22 cows per hectare. Thus a major challenge to grassland research and development is to show how to produce sufficient high-quality feed from grass and forage crops to support a yield of 6000 litres of milk at a concentrate input of 1 tonne per cow. A further challenge is to demonstrate that any reduction in stocking rate need not be reflected in a reduction in total farm revenue.

Milk from grass alone

Concentrates are normally credited with a metabolisable energy value of 11 MJ per kg (12 MJ per kg of dry-matter). Cows given concentrates *ad lib* with minimal essential fibre to maintain rumen function, would be expected to yield well over 8000 litres of milk in a lactation.

Table 9.5   Milk from grass: targets for profitable production*.

Milk yield (litres/cow)	6000
Concentrates (kg/cow)	925
Nitrogen (kg/ha)	370
Stocking rate (cows/ha)	2.05
UME (GJ/ha)	93
Gross margin	
(£/cow)	586
(£/ha)	1201

*Average grass growing conditions, autumn calving herd.

Plate 9.1 A lactation yield of 4700 litres may be expected from cows given grass alone.

By contrast, preliminary trials at ICI's research station at Jeallot's Hill, Bracknell, Berkshire, show that with cows calving in late January, a lactation yield of 4700 litres might be expected from grass alone (see table 9.6). The silage offered as the sole feed (with supplementary minerals and vitamins) had a metabolisable energy content of 11.2 MJ per kg of dry matter (DM). The cows were turned out to pasture after receiving silage for the first 105 days of lactation and it

Table 9.6   Milk from grass alone: the Jeallot's Hill pilot trial.

Number of animals	18
Milk yield* (litres/cow)	4705
Concentrates	zero
Nitrogen (kg/ha)	450
Stocking rate (cows/ha)	2.3

* Average calving, late January.
Silage of 11.2 MJ metabolisable energy per kg dry-matter was given as the sole winter feed, followed by grazed pasture.
Average daily milk yield from silage was 21.4 kg per day.

is unlikely that they would have eaten grass with an ME content much lower than 12 MJ per kg DM. Thus the major difference between concentrates and high quality grass as feeds for productive dairy cows lies in the amount which can be consumed per day.

If high-quality grass alone can support a 'herd' lactation yield (including heifer lactations) — albeit on a pilot scale — which is only slightly below the national average of 5100 litres, then a key area for future development is the production of concentrate feeds which complement the attributes of grass and which will lead to increases in total nutrient supply. Specifically, the most useful supplements for high-yielding cows are those which will increase the total supply of ME and essential amino acids to the body tissues. This means that concentrates should have a high content of both ME and undegraded dietary protein (UDP), and, to avoid high rates of substitution, should be given in relatively low total amounts per feed.

Summer milk production

Other things being equal (which they are not) the most profitable system of milk production from grass would rely almost entirely on grazed pasture because it is relatively cheap and of relatively high quality. In practice, however, autumn calving predominates because it is more profitable than spring calving and summer milk production. The milk processing industry would be unable to cope with the glut of milk in early summer if there was a marked swing towards spring calving. But for the producer-processor, there may well be a case for having a proportion of spring-calving cows in the herd.

At Trawsgoed Experimental Husbandry Farm, and at ICI's Ravenscroft Farm, cows calve in January and February and are given a flat rate of concentrates until turnout to pasture. For three winters at Ravenscroft, the response of the cows to two levels of concentrates was compared. The results are given in table 9.7. Using silage of only moderate quality, the overall response to the higher level of concentrate, at 1.8 kg milk per kg extra concentrate dry matter was economic. The rate of substitution of silage DM by concentrate DM was relatively low (0.4), reflecting the quality of the silage.

An interesting feature of the trials is that the direct response to extra concentrate during the winter period, 0.8 kg milk per kg concentrate DM, was uneconomic. But residual effects during the much longer grazing season, which occurred irrespective of grazing pressure, meant that overall the extra concentrate input was worth-

Table 9.7 Concentrates for spring-calving cows*.

	Low	High
Concentrate (kg/cow)	430	750
Milk yield (litres/cow)	4684	5183
Stocking rate (cows/ha)	2.4	2.2
UME (GJ/ha)	105	101
Margin over feed and forage (£/ha)	1084	1212

* Average results from 3 trials, 1980 to 1982.
With silage of 9.8 MJ ME per kg dry matter, the overall response to extra
concentrate (1.8 kg milk/kg concentrate dry matter) was economic.

while. The average margin over feed and forage, for the higher level
of concentrate, was comparable to the autumn-calving herds of the
top grassland farmers in the Rex Paterson Memorial Study.

Legumes

Legumes play a minor role in the UK dairy industry, though both
lucerne and red clover have received attention from research workers
in recent years. White clover was neglected until recently, but a series
of detailed studies at the Grassland Research Institute, including
trials with dairy cows, has shown the potential of the crop as a feed
for productive animals. The results of a trial with spring-calving cows
at pasture are shown in table 9.8.

Table 9.8 White clover for milk production.

	Perennial ryegrass	White clover
Daily milk yield* (litres/cow)	22.2	25.0
Milk composition		
protein (%)	2.96	3.11
fat (%)	4.15	3.89
lactose (%)	4.94	4.97

* Average, weeks 3 to 18 of lactation, at pasture with no supplementary feeds.
Cows ate more dry matter and gave more milk when grazing white clover than
when given ryegrass.

Cows grazing white clover ate more dry matter than those given ryegrass, and during the grazing season averaged 13 per cent more milk per day. After the grazing period all cows received a standard diet based on grass silage. The differences in yield seen during the grazing season persisted through the lactation, so that total lactation yields averaged 5900 and 5100 litres for cows grazed on white clover and grass, respectively.

The milk from both groups of cows was tested at the National Institute for Research in Dairying for its flavour, composition and processing properties. There were only slight differences in composition (see table 9.8), and the milk from cows given clover was similar in flavour to that from cows given ryegrass. But the clover pasture gave rise to milk of higher casein content which produced firmer curds following the addition of rennet.

In practice, cows are more likely to receive mixtures of grass and clover rather than pure swards of clover, and the response of cows to such mixtures is now being studied. Significantly, the higher feed intake characteristics of the pure clover sward appear to remain when cows are offered mixed clover/grass swards. This offers the exciting possibility that in the wetter areas of the country where grass grows well, it may be possible to achieve acceptable herbage yields, high feed intakes and high yields of milk from swards which contain, say, 30 per cent of the dry matter as white clover and which receive little or no fertiliser nitrogen. A further advantage is that the milk may have improved cheese-making characteristics compared to the milk from conventional grass pastures.

In Brittany, farmers using grass/clover for summer grazing are being recorded by the Institute Technique d'Élevage Bovin (ITEB). Some forty farmers are being surveyed and the results show that the average level of performance is only slightly lower than from pure grass swards given 250 to 300 kg nitrogen per hectare. However, the range of performance is very large and closely related to the proportion of clover in the sward. Targets are now being set, based on the achievements of the best farmers who have managed to maintain more than 30 per cent of clover in their pastures.

The performance of two of the top herds is summarised in table 9.9. Significantly, the farms also grow high-energy forages for winter feeding, and give their cows a daily forage supplement during the grazing season. On one farm, hay is given at 1 kg/day; on the other maize silage is given at 3 kg/day throughout the summer. Clearly the use of fodder beet and maize has enabled a relatively high stocking rate and UME output from forages to be achieved. The grass/clover

Table 9.9 Milk from grass/clover swards in Brittany.

	Farm A	Farm B
Winter feeds	Hay, fodder beet	Maize silage
Summer feeds	grass/white clover	grass/white clover
Milk yield (litres/cow)	6300	5800
Concentrates (t/cow)	1.1	0.9
Nitrogen fertiliser on grass (kg/ha)	30	50
Stocking rate (cows/ha)	2.4	2.2
UME (GJ/ha)	111	101

swards receive very low inputs of fertiliser nitrogen, usually one application in early spring.

Future prospects

Looking further ahead, the most probable trend is towards a New Zealand type of dairying based on grass/clover mixtures in areas where grass grows well, and a Dutch type of dairying based on maize and lucerne in the drier parts of the country. In both cases the emphasis will increasingly be on efficiency rather than on output. Improved techniques for establishing and managing legume crops should allow economies in fertiliser nitrogen to be achieved. But it will probably take a 30 per cent rise in the relative cost of fertiliser nitrogen to produce a significant move towards white clover. In addition, the possibility exists that the genes for nitrogen fixation may eventually be introduced into ryegrasses and forage maize, though at a penalty to yield. The fixation of nitrogen by the bacteria in the root nodules occurs only at an energy cost to the plant, thus imposing a constraint on the maximum rate of plant growth. A further possibility is that legume varieties will be developed which do not cause bloat. This achievement alone will be a major breakthrough in removing a problem associated with grazed legumes.

There is clearly enormous scope for increasing the efficiency with which grass and forage crops are used for the production of milk. In exceptional years, such as 1976 and 1983, the best laid plans can be totally confounded by the weather. In good years for grass production, therefore, we should be looking to conserve the excess as a buffer for leaner years. Alternatively, we should perhaps hedge our bets by growing forage crops such as maize which complement conventional grasses, and which grow particularly well in dry summers when grass yields are low.

References

1. Amies, S.J. (1983) *Farm Management Services Report No. 37*, MMB.
2. Walsh, A. (1982) *The Rex Paterson Memorial Study*, British Grassland Society.
3. Thomas, C. and Young, J.W.O. (1982) *Milk from Grass*, ICI/GRI.

10 Profitable Beef from Grass

It was timely that the Money from Grass '83 campaign, with its aim of creating a greater awareness of the potential of grass as a feed for productive cattle and sheep, was held exactly twenty years after the establishment of the Beef Recording Association. Before that time there was no systematic scheme for recording the performance of cattle on commercial farms, and the concept of planned systems of production was very much in the minds of the boffins rather than a reality on the farm.

The pioneering work of the BRA was recognised, together with that of the Pig Industry Development Association, when the Meat and Livestock Commission was formed in 1968. Since then there has been a steady development of livestock recording services. Those producers who have kept in close touch with the Commission's advice have had the benefit of an increasingly comprehensive service.

The 18-month system of beef production

In the early days of the BRA, one system of beef production emerged as having greater potential than the others. It involved the rearing of autumn-born, dairy-bred calves to a slaughter weight of 475 kg at 18 months of age. The system had the advantage of a lower reliance on cereal grain than the barley beef system, yet it allowed greater numbers of cattle to be kept on each hectare of land than those systems which involved slaughter at two or more years of age.

Fenwick Jackson, who farms near Berwick-on-Tweed, is one of the pioneers of the 18-month system. He was one of the eight winners of the 1983 'Grass to Meat' awards. Very high stocking rates coupled with excellent stockmanship enabled Mr Jackson to produce 1800 kg of liveweight per hectare in 1982/83. The system operated by Mr Jackson has evolved over the last twenty years. The aim is to speed up the rate of growth and reach slaughter weight as early as possible. Many of the calves are now kept entire as bulls. Daily gain at grass averaged 1.14 kg in 1982.

Gross margins

The extent to which the 18-month grass/cereal system of beef production has produced acceptable margins over the five-year period from 1978–82 is shown in table 10.1 compared to other systems of production. The data are from the Meat and Livestock Commission.[1] Gross margins per head, adjusted for inflation, were slightly higher for the 20/24-month system, principally because of the higher weight at slaughter of the older animal. But the 18-month system showed a substantial advantage in margin per hectare over the other systems.

Table 10.1 Beef from grass: gross margins 1978–82, five-year averages adjusted for inflation.

System	Gross margin per head (£)	Gross margin per hectare (£)
Suckled calf production		
lowland herds	149	283
upland herds	186	223
Dairy-bred calves		
18-month beef	196	613
20/24-month beef*	210	443
Store cattle		
winter finishing	62	—
grass finishing	59	330

* 1979–81 results.
The 18-month grass/cereal system has consistently given the highest margins per hectare.

'Grass to Meat' award winners

Success in generating high margins in suckled calf and in finished beef production means having a tight stocking rate. In the case of suckled calf production (see table 10.2) it also means producing a high weight of calf at weaning. The six suckled calf producers who won the 'Grass to Meat' award over the five years 1979–1983, all had a stocking rate in excess of 1.8 cows per hectare. The three lowland herds averaged 2.4 cows per hectare. Average output of weaned calf was over 300 kg per cow.

The eight award-winning 18-month beef systems achieved their

Table 10.2   Suckled beef production: award winners compared to average.

	Average	Award winners
Stocking rate (cows/ha)	1.5	2.1
Liveweight gain (kg/calf per day)	0.9	0.96
Liveweight output (kg/ha)	361	698
Gross margin* (£/cow)	168	198
(£/ha)	253	410

* Five-year averages, adjusted for inflation to 1981/82 prices.
Award winners had a higher stocking rate and produced a heavier weight of calf
at sale. Gross margin per hectare was 62% higher for award winners than for the
average recorded herd.

success by combining high rates of gain at pasture with high stocking
rates (see table 10.3). Output from grazing exceeded 1000 kg of liveweight gain per hectare, and although gross margin per head was only
slightly higher than average, gross margin per hectare, at almost £950
was exceptionally high. Incidentally, the comparable gross margin
per hectare of forage from BOCMS-costed dairy farms was £920 over
the same period.

Targets

Targets for the performance of beef cattle are shown in table 10.4
for a range of systems of production. The emphasis is on achieving

Table 10.3   18-month beef: award winners compared to average.

	Average	Award winners
Stocking rate (cattle/ha)	3.1	4.8
Liveweight gain (kg/day)	0.73	0.83
Grazing gain (kg/ha)	776	1156
Gross margin* (£/head)	196	220
(£/ha)	613	948

* Five-year average, adjusted for inflation to 1981/82 prices.
Award winners achieved better performance at pasture, coupled with high
overall stocking rate, to realise a gross margin per hectare which was 55% higher
than the average.

high rates of gain from grazed and high-quality conserved pasture. Concentrate feeds are limited to the post-weaning period in the case of suckled calves, and to the autumn grazing and indoor-feeding periods in the case of dairy-bred and store cattle.

Table 10.4 Beef from grass: targets for performance.

System	Daily gain (kg)	Liveweight at weaning or slaughter (kg)	Concentrates (kg)	Stocking rate (cattle/ha)
Suckled calves*	1.0	300	0.2	2
Dairy-bred calves				
18-month beef	0.9	500	0.7	4
24-month beef	0.8	550	0.5	3
silage beef	1.0	500	0.5	6
Store cattle†				
winter finishing	0.8	475	0.5	—
summer finishing	0.9	450	0.2	4

* Autumn-calving upland herds, large breed of sire.
† British crossbred steers.

Stocking rate is a rather vague term which takes little account of the conditions of the individual farm — its soil and its grass — or of the variation in rate of grass growth during the season, or of the level of nitrogen fertiliser input. For those farmers who weigh their cattle at turnout and during the grazing season, the stocking rate should bear some relationship to the total weight of live animal per hectare and the level of fertiliser nitrogen used. Suitable targets for beef units with either the 18-month system or store cattle on lowland grass are an average stocking rate during the grazing season of 2000 kg liveweight per hectare (i.e. four cattle weighing 250 kg liveweight) and 300 kg nitrogen fertiliser per hectare (see table 10.5). These targets are more appropriate to leys than to permanent pasture since there is evidence that leys can respond better to high levels of fertiliser than permanent swards.

Grass/clover swards

An alternative for farmers with permanent grass, is to rely on grass/clover swards with limited use of fertiliser nitrogen. The results of a

Table 10.5   Beef from grass: targets for stocking rate and output.

Stocking rate (kg liveweight/ha)	
May and June	2500
July and August	2000
September and October	1500
Output (kg liveweight gain/ha)	1000

Two tonnes of liveweight per hectare stocked on land receiving
300 kg fertiliser-N per hectare should generate one tonne of
liveweight gain.

comprehensive trial at Greenmount and Loughry Agricultural Colleges
in Northern Ireland[2] to study the relative merits of high or low levels
of nitrogen use with grass/clover swards in 18-month beef production, are summarised in table 10.6. Clover-dominant swards proved
no more difficult to manage than high-N swards, but strong grass
growth in 1980 followed by a wet winter and cold wet spring in
1981 seriously reduced clover growth.

Animal performance was better when there was a high content of
clover in the swards. Contents of over 30 per cent of clover in the
dry matter are necessary during the summer months to stimulate
better liveweight gains.

The stocking rates were chosen to match the likely yields of grass.
They averaged 1000 kg liveweight per hectare on the grass/clover,
low-N swards and 1200 kg liveweight per hectare on the high-N
swards. Liveweight output was 24 per cent higher from the swards
which received the higher fertiliser nitrogen (see table 10.6).

A striking feature of the trial was that despite the lower output,

Table 10.6   Beef from grass/clover swards.

	High N (300 kg N/ha)	Low N (50 kg N/ha)
Daily gain		
at pasture	0.84	0.91
overall	0.77	0.82
Liveweight output (kg/ha)	1049	847
Gross margin (£/ha)	565	530
Working capital (£/ha)	1256	896
Return on capital (%)	45	59

the low-N system proved to be almost as profitable as the high-N system – at least in terms of gross margin per hectare. Further, the low-N system required 30 per cent less working capital per hectare, and return on capital was higher for the low-N system than for the high-N system.

Inputs of N approaching 300 kg per hectare are rarely encountered in grass beef enterprises. Thus there is a strong case for encouraging a greater dependence on clover and the strategic use of N to boost grass growth in periods of grass shortage.

A challenge to farmers and research workers is to devise management practices which will maintain a high proportion of clover in the sward year after year. One of the major problems encountered in the Northern Ireland trial was the variable contribution from clover in the low-N system over the six years of the trial.

Feedlot beef

At the other extreme, farmers with land suitable for silage production, good buildings and capital available to fill them with cattle may well wish to consider a feedlot system in which the cattle do not graze at all. This system, which has been widely adopted in countries where forage maize is grown, is particularly well suited to bull beef production. In the UK, the system developed at Rosemaund Experimental Husbandry Farm, which is based on grass silage, has now been tested at the Beef Unit of the National Agricultural Centre, and on a small number of commercial farms. The results are summarised in table 10.7.

Table 10.7   Feedlot beef from silage.

	Grass silage	Maize silage
Daily gain (kg)	1.0	1.0
Feeding period (months)	14	12
Feed: silage (t/head)	5.0	5.8
concentrates (t/head)	0.93	0.85
Stocking rate (cattle/ha)	7.3	6.6
Liveweight output (kg/ha)	2600	2370
Gross margin (£/head)	169	160
(£/ha)	1227	1052

Initial results showed that under commercial conditions high levels of output were obtained from these intensive systems.

The average input of concentrates, at just under a tonne per head, is comparable to that used in the 18-month system. With grass silage the cattle receive 2 to 3 kg concentrates per day, with the level of supplementation increasing towards the end of the feeding period. In the maize silage system, a flat rate of 1.5 kg concentrates per day can be given. The composition of the supplement is varied during the feeding period to take account of the low content of protein in the silage and the reducing requirement of the animal for protein as it become heavier.

The feedlot system obviously requires greater storage capacity for silage, and additional machinery may be required to conserve the crop and feed the animals. Stocking rates, however, are substantially higher than in those systems which rely on grazing. Gross margins per hectare are also relatively high.

Feedlot beef clearly holds promise for farmers with land suited to forage conservation and buildings suited to feeding cattle throughout the year. The system is well suited to the continuous production of finished cattle of specified weight and fatness. It eliminates the

Plate 10.1 Feedlot beef from grass silage or from forage maize is particularly well suited to farms with suitable buildings and enough capital to fill them with cattle.

complicated problems of balancing unpredictable grass growth with the increasing feed requirements of the grazing animal.

Lucerne silage

In France, where feedlot beef has been produced for many years, there is increasing interest in using lucerne silage in combination with maize silage for bulls. The results of a recent trial at the Institute Technique des Céréales et Fourrages[3] (table 10.8) illustrate the very high rates of daily gain which were recorded from a diet of well-preserved lucerne silage and limited concentrates. Lucerne was the sole forage feed until the bulls reached 430 kg liveweight, when maize silage was introduced into the diet for the final finishing period. Rapid growth was sustained so that when slaughtered at the end of the 11½-month feeding period, the Normand bulls weighed over 600 kg liveweight. The trial demonstrated that it was possible to produce carcases of 330 to 340 kg from lucerne silage supplemented with 2 kg per day of concentrate up to 430 kg liveweight, and from maize silage supplemented with 1.3 kg of concentrate per day thereafter.

Selecting cattle for slaughter

The target slaughter weights in table 10.4 refer to the major systems of beef production. They do not take into account the fact that

Table 10.8   Feedlot beef from lucerne silage.

Liveweight at start (kg)	170
Liveweight at slaughter (kg)	605
Daily gain (kg)	1.26
Feeding period (months)	11.5
Feed	
Lucerne silage (t at 22% DM)	5.6
Maize silage (t at 32% DM)	3.7
Concentrates (t at 85% DM)	0.75

The very high rate of gain was maintained during both the growing phase, when the bulls were given lucerne silage as the sole forage, and also during the finishing phase when they were given maize silage. Overall, lucerne and maize silage each accounted for 40% of the total dry-matter consumed.

Plate 10.2 Lucerne for silage; very high levels of daily gain have been achieved from bulls given lucerne and maize silage with limited concentrates.

there are differences in rate of finishing according to breed type and sex of the animal. For example, a large-framed Charolais cross will tend to reach the same carcase fatness as an Angus cross calf at a considerably heavier weight. At the same age, therefore, the larger-framed animal will be both heavier and leaner. These differences are illustrated in fig. 10.1 for two systems of production, both of which use calves born in dairy herds.

When selecting cattle for slaughter, the target weights at specified ages should be used as the main criterion. This is because cattle move only slowly between the different carcase fat classes; it may take up to six weeks for a change to occur.

Conformation is an additional consideration to take into account. Cattle of poor shape have lower saleable meat yields in their carcases. But it is usually not cost-effective to attempt to improve an animal with poor conformation, but with adequate carcase fatness, by excessive finishing (see checklist). The general point to remember is that an over-fat animal is costly to produce and has a reduced saleable meat yield (table 10.9).[4]

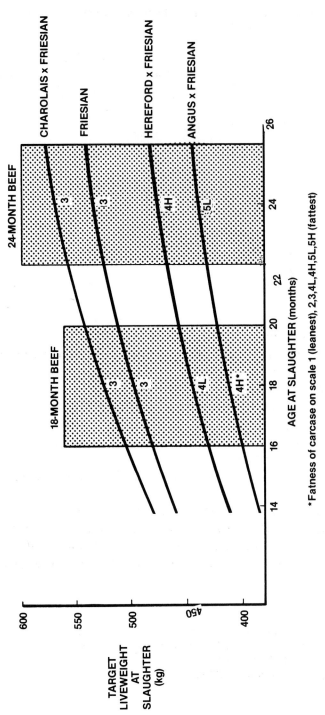

*Fatness of carcase on scale 1 (leanest), 2,3,4L,4H,5L,5H (fattest)

Fig. 10.1 Target weights at slaughter and probable carcase fatness for cattle of different breeds.

Table 10.9 Saleable meat yield in relation to carcase fatness and conformation (% of carcase weight).

		Fat Class					
		(Leanest)					(Fattest)
		1 and 2	3	4L	4H	5L	5H
(Very good)	E and U+			72.4			
	U			71.7			
Conformation	R	74.5	72.5	71.0	70.0	69.0	66.5
Class	O			70.3			
	O–			69.6			
(Very poor)	P			68.9			

A checklist for selecting finished cattle for slaughter
1. Use target weights for the type of animal and system of production (see fig. 10.1).
2. Expect variation in a group.
3. Expect each animal to remain in the same fat class for up to six weeks.
4. Select individual cattle on the basis of their weight, breed type and condition (or conformation).
5. Check weights and condition against likely carcase fatness (fig. 10.1).
6. Accept conformation as a characteristic of breed type.
7. Avoid attempting to improve conformation by excessive finishing.
8. Select an optimum finish (fat class) for market requirements.
9. Remember: excess fat costs more in feed and reduces meat yield.

References

1. MLC (1983) *Beef Yearbook 1982/83.*
2. Stewart, T.A. *et al. Agriculture in Northern Ireland* **58**, Nos 2 and 3.
3. ITCF (1983) *Annual Report*, 43.
4. MLC (1983) *Beef Yearbook 1982/83.*

11 Profitable Lamb from Grass

Lamb, for many years the poor relation of the livestock industry, has received considerably more attention from the farmer and adviser since the introduction of the EEC Sheepmeat Regime in October 1980. Two notable effects of the variable premium scheme for lamb and the ewe premium were firstly that margins increased in 1981 by more than the rate of inflation. Secondly, the way the variable premium scheme was structured focused the minds of farmers on the earlier marketing of finished lambs. Delay beyond August in selling finished lambs meant lower total returns. The net effect of the trend to earlier marketing has been a sharp fall in market prices in June, July and August. Consequently, high levels of variable premium were paid to producers during this period of the year.

The high proportion of total returns from the premium when most lambs are sold makes it essential that the lambs qualify for premium payment. This means that lambs tend to be marketed over-fat rather than under-finished. Thus the design of the premium scheme is at variance with market requirements. Perhaps there should be graded rates of payment based on carcase classification with the fattest lambs (i.e. class 4 and 5) receiving nothing, and the highest payments for carcases in fat classes 2 and 3L, which the meat trade requires.

From the point of view of land use and the conservation of wild-life and landscape features, sheep are generally considered to be 'good'. This is particularly true in the east and south-east of the country, where the grazing of sheep is associated with grass, hedges and an increase in feed supply for birds. Thus the sheep farmer is likely to continue to receive support from the conservationist and environmental lobbies.

Gross margins

Whilst not yet showing comparable margins per hectare to those generated by the 18-month or 24-month beef production system, the average gross margins in recorded[1] lowland and upland flocks over the five-year period 1977–81 (table 11.1) were very comparable to those of suckled calf production and store cattle finishing. Early lambing flocks tended to be kept at higher stocking rates, and in consequence they produced higher margins per hectare than those generated by flocks which lambed in March and April (see table 11.1).

Table 11.1 Lambs from grass: gross margins, 1977–81, five-year averages adjusted for inflation.

System	Gross margin per ewe (£)	Gross margin per hectare (£)
Lowland flocks lambing in:		
December/January	27.8	350
March/April	28.5	308
Upland flocks	32.1	238
Hill flocks	23.4	—

Early lambing of lowland ewes was reflected in similar margins per ewe, but because overall stocking rates were higher, margins per hectare were also higher.

'Grass to Meat' award winners

Many of the 'Grass to Meat' award winners have been enthusiastic flockmasters who achieved very high levels of output from intensively managed grass. One of the 1983 winners, John Coultrip, produced almost 1500 kg of lamb liveweight per hectare. His grassland management policy is the same as if it was for a dairy herd. Mr Coultrip stocks his Romney sheep at 25.8 ewes per hectare, and realised a gross margin of £951 per hectare in 1982. With 1.52 lambs reared per ewe, there is no evidence that individual ewe performance was adversely affected by the high stocking rate. Flushing the ewe is vitally important, to take advantage of the twinning abilities of the rams.

On average, the 'Grass to Meat' award winners reared 0.2 lambs per ewe more than the average of MLC flocks at a higher stocking rate than the average. Gross margin per hectare was 66 per cent higher than the average (see table 11.2).

Table 11.2   Lamb from grass: award winners compared to average.

	Average	Award winners
Output (lambs reared per ewe)	1.4	1.6
Stocking rate (ewes/ha)	13	16
Liveweight output (kg/ha)	640	1007
Gross margin* (£/ewe)	28	40
(£/ha)	329	546

* Five-year average, adjusted for inflation to 1981/82 prices.
Award winners reared more lambs per ewe at a higher stocking rate than the average, so that total output of liveweight exceeded 1000 kg a hectare.

Targets for performance

Targets for performance in lowland, upland and hill flocks are shown in table 11.3. The emphasis should be on producing as high a proportion of finished lambs as possible. The target levels of concentrate feeding take account of the need to feed the ewe a high-energy diet in late pregnancy, and to give the lambs access to concentrates after weaning to maintain high rates of growth until they are sold as finished lambs.

Plate 11.1 The emphasis in lamb production should be on producing as high a proportion of finished lambs as possible. Silage can be a useful supplement to autumn pasture for finishing lambs off grass.

Table 11.3 Lamb from grass: performance targets.

System	Lambs reared/ewe	% sold finished off grass	Concentrates (kg/head)		Stocking rate (ewes/ha)
			ewe	lamb	
Lowland flocks lambing in:					
December/January	1.5	90	50	30	18
March/April	1.5	80	40	5	16*
Upland flocks	1.4	60	30	5	16*
Hill flocks	1.2	30	20	5	—

* Medium or larger breed in lowland flocks, smaller type in upland flocks.

In lowland flocks, stocking rate is the most important factor contributing to higher margins. The top third of MLC-recorded flocks not only achieved a higher output per ewe, they did so at a higher stocking rate. They also applied more fertiliser nitrogen and stocked their land in relation to their input of fertiliser. Detailed records taken by the Meat and Livestock Commission[2] illustrate the higher weight of sheep stocked at the higher levels of nitrogen use (see table 11.4).

Top hill flocks had a higher percentage of fertile ewes (93 per cent compared to an average of 91 per cent). They also sold a higher proportion of draft ewes, indicating that they were applying a higher selection pressure on their flocks through culling. The top third retained a higher-than-average proportion of the lamb flock on the farm for feeding or for breeding. Not surprisingly they sold a higher proportion of their lambs finished off grass. The targets for hill flocks are summarised in table 11.5.

Table 11.4 Fertiliser nitrogen and stocking rates in recorded lowland sheep flocks.

n fertiliser-N (kg/ha)	Stocking rate (kg liveweight/ha)
60–120	930
120–180	1050
180–240	1200
over 240	1450

Table 11.5   Targets for hill flocks.

Ewes	
Percentage lambed	93
Sold as draft ewes (%)	25
Lambs (per 100 ewes to the ram)	
Reared	120
Sold finished off grass	40
Sold or retained for feeding	45
Retained for breeding	35

Improvement of hill land, outlined in chapter 4, is often accompanied by breed substitution whereby a small hill breed is replaced by a larger-framed type which is capable of rearing more lambs. Thus the prospects are that selection within, as well as between breeds of hill sheep will continue to place emphasis on weight of weaned lamb per ewe as a major factor. This in turn will lead to the selection of larger-framed animals which will effectively increase stocking weight.

Unfortunately, subsidies are paid to hill sheep farmers on a headage basis. This procedure encourages the retention of the smaller animal and exacerbates the problem of the excessively fat lamb which does not fit easily into the specification now required by meat wholesalers and retailers.

Patterns of production

The patterns of lamb sales from MLC-recorded flocks[3] are shown in fig 11.1 for a range of systems. Early lambing flocks selling finished lambs had sold almost 80 per cent of their lambs by the end of September. Those that were sold in October were either store lambs or lambs for breeding. The price structure favours the selling of lambs earlier in the summer, rather than in August or September. But store lamb feeders chose to retain over half their lambs for winter finishing when both the price, and the weight of finished lamb, are higher.

Selecting lambs for slaughter

If the variable premium scheme were to be adjusted so that those

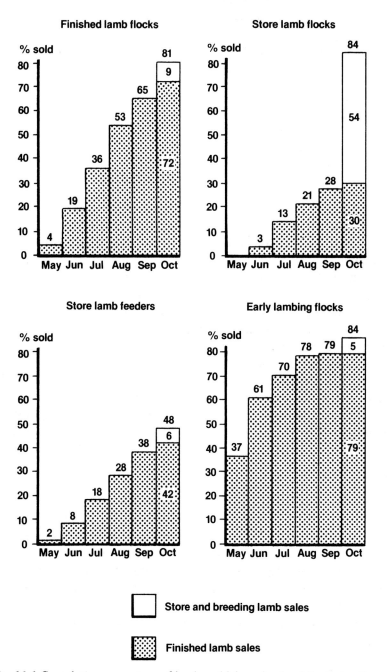

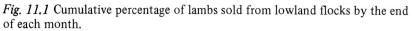

Fig. 11.1 Cumulative percentage of lambs sold from lowland flocks by the end of each month.

carcases which meet market requirements receive support while the rest do not, then the accurate selection of lambs for slaughter based on carcase fatness would be an even more important determinant of economic success. Unfortunately there is no satisfactory way of objectively assessing carcase composition in the live animal. The development of a simple method of so doing would be welcomed by many in the industry. Nevertheless, lambs are usually inspected at regular intervals to estimate carcase fatness before being sold.

At Trawsgoed Experimental Husbandry Farm[4] lamb selection is based on handling along the spinous processes of the backbone and over the tips of the transverse processes of the lumbar vertebrae. The shoulder area is also considered. Regular handling of each group of lambs ensures that the proportion of lambs which are sold either too thin or too fat is kept to a minimum (see fig. 11.2). This is particularly relevant when lambs are sold deadweight to a contract specification. The contract usually includes a penalty for lambs which are over-fat (MLC Class 4).

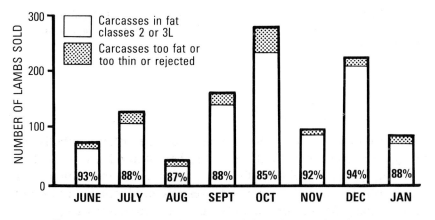

Between 85 and 94% of lambs sold from Trawsgoed EHF in 1982 were in the target MLC fat classes of 2 and 3L.
Average carcass weight was below target in one month only (August). Regular handling of lambs is essential to achieve the targets for carcass weight and composition

Fig. 11.2 Matching lamb production to market requirements.

The target lamb

The concept of the target lamb is now accepted by most authorities in the sheep industry. The target lamb has a carcase weight of 16 to 19 kg, and a carcase fatness of either 2 or 3L. In the case of Trawsgoed EHF, the average carcase weight was below target in only one month of 1982.

Plate 11.2 Regular inspection of lambs is necessary to ensure that a high propor-tion meet the target specification of 16 to 19 kg carcase weight and 2 or 3L carcase fatness.

Clover for finishing lambs

Grass accounts for about 90 per cent of the total metabolisable energy (ME) consumed by the ewe and her lambs. Thus the cost of producing grass, principally fertiliser, is relatively more important than in other systems of ruminant livestock production. In lowland spring-lambing flocks, for example, fertiliser represented 26 per cent of total variable costs.

It might be expected, therefore, that grass/clover swards might offer a considerable saving in costs to the sheep farmer, compared to grass swards alone. Unfortunately sheep have a preference for clover in mixed swards, and at relatively high stocking rates it is difficult to prevent clover from being eaten out of the sward. A more feasible proposition, therefore, is to grow clover-dominant swards for finishing the lambs after weaning. The progressive removal of finished lambs from the pasture should help the clover to survive.

A very important feature of such a strategy is that maximum lamb growth is achieved on clover swards at only half the daily pasture allowance of that of ryegrass. Further, the well-known superiority of the legume over grass in improved lamb growth rate was also apparent at lower levels of pasture allowance in trials at Ruakura Agricultural Research Centre, New Zealand (see fig. 11.3).[5] This suggests that a much smaller area of special-purpose legume pasture is needed than is the case with grass, despite the possibility that the yield of the legume may be somewhat less than that of well-fertilised grass. The prospect of weaned lambs gaining weight at 200 g a day should be attractive enough to encourage farmers to investigate the feasibility of sowing a special area of clover for finishing lambs off pasture at low cost.

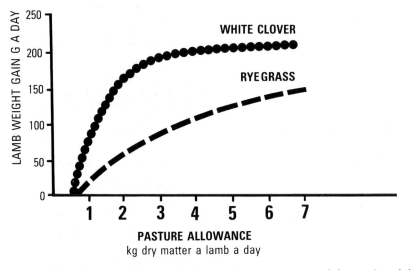

Maximum liveweight gains were achieved with clover at only half the pasture allowance of that of ryegrass

Fig. 11.3 Finishing lambs on legume pastures.

References

1. MLC (1982) *Commercial Sheep Production Yearbook 1981/82.*
2. MLC (1983) *Sheep Yearbook.*
3. *Ibid.*
4. Griffiths, M.S. (1983) *Occasional Publication No. 8*, British Society of Animal Production.
5. Jagush, K.T. *et al.* (1979) *Proceedings of the 31st Ruakura Farmers' Conference*, Hamilton, New Zealand, 47—52.

Index